5TH EDITION

PHYSICAL GEOGRAPHY LABORATORY MANUAL

DALTON W. MILLER, JR.
ANDREW E. MERCER

MISSISSIPPI STATE UNIVERSITY

New York   Oxford
OXFORD UNIVERSITY PRESS

Oxford University Press is a department of the University of Oxford.
It furthers the University's objective of excellence in research,
scholarship, and education by publishing worldwide.

Oxford   New York
Auckland   Cape Town   Dar es Salaam   Hong Kong   Karachi
Kuala Lumpur   Madrid   Melbourne   Mexico City   Nairobi
New Delhi   Shanghai   Taipei   Toronto

With offices in
Argentina   Austria   Brazil   Chile   Czech Republic   France   Greece
Guatemala   Hungary   Italy   Japan   Poland   Portugal   Singapore
South Korea   Switzerland   Thailand   Turkey   Ukraine   Vietnam

For titles covered by Section 112 of the US Higher Education
Opportunity Act, please visit www.oup.com/us/he for the
latest information about pricing and alternate formats.

Published by Oxford University Press
198 Madison Avenue, New York, NY 10016
http://www.oup.com

ISBN 978-0-19-024687-7

Printing number: 9 8 7 6 5 4 3 2 1

Printed in the United States of America
on acid-free paper

CONTENTS

MAPPING THE EARTH'S SURFACE

GOALS OF THIS LAB:

To correctly identify locations on a map and on a globe using latitude and longitude. To identify the type of map projection and the map scale type being used in your atlas.

REQUIRED ITEMS FOR LAB:

Textbook, atlas of the United States, a globe, and a computer with Internet access (for hi-resolution images)

We use maps daily to identify our location or the location of a place or landmark that we are seeking. Maps are used to represent a large, three-dimensional world into a two-dimensional depiction of that world. The amount of information that the map represents is directly related to the *scale* of the map. The scale of a map is the ratio of the size of an object on a map to the actual size of the object that the map represents. Map scales are often represented as ratios comparing the distance on a map with the distance in the region that was mapped. For example, a scale of 1:100,000 means that 1 cm on a map represents 100,000 cm on Earth, or 1 km.

Maps often represent position by overlaying a grid on a two- or three-dimensional surface. For Earth, this is a spherical grid (since the Earth is roughly spherical) and consists of lines that extend east–west (*latitude lines—parallels*) and lines that extend north–south (*longitude lines—meridians*). A *parallel* measures north–south distance from a standard latitude (the equator is at 0° latitude) and extends from 90°N (the North Pole) to 90°S (the South Pole). Latitude lines are called *parallels* because the lines are parallel to each other (Fig. 1.1A). A *meridian* measures distance in the east–west direction from a standard longitude line (the Prime Meridian is at 0° longitude) and extends from 180°W to 180°E. The 180° west–east longitude line is denoted as the *International Date Line*. Meridians converge on the North Pole and South Pole and are farthest apart near the equator (Fig. 1.1B).

As expected, when trying to represent a spherical Earth on a two-dimensional surface, errors are unavoidable. Cartographers use many different methods to make this conversion, called *map projections*. A map projection is defined as an orderly arrangement of meridians and parallels produced by any systematic method that can be used for drawing a map of a spherical Earth on a flat surface. Map projections attempt to minimize the errors associated with representing a spherical Earth on a two-dimensional map, but often have to sacrifice either the correct size of a feature on the map (*equal area projection*) or the correct shape of a feature (*conformal projection*).

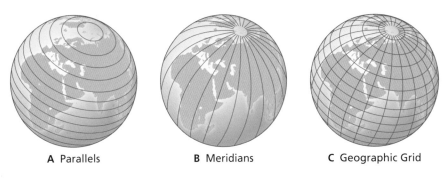

A Parallels **B** Meridians **C** Geographic Grid

Figure 1.1 Parallels (A) and meridians (B) on a globe. Parallels run east–west; meridians run north–south.

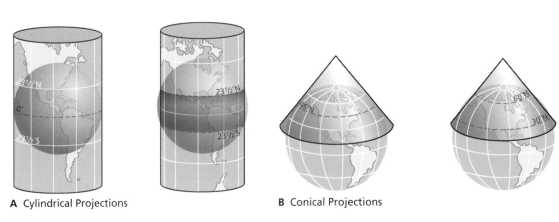

A Cylindrical Projections **B** Conical Projections

Figure 1.2 Construction of cylindrical (A) and conic (B) projections with one and two standard parallels.

Many different types of map projections are commonly used in cartography. One of the most famous types is a *cylindrical projection*, where the Earth's latitude–longitude grid is transferred to a cylinder that is then cut and laid flat (Fig. 1.2A) such that the center of the cylinder passes through the Earth's axis of rotation. When finished, the latitudes and longitudes appear as perpendicular straight lines. The image on a cylindrical projection becomes more distorted the farther from the *standard parallel* you look, but at the standard parallel, or the latitude line that is tangent to the globe, no distortion is seen. The most famous cylindrical projection is the *Mercator projection*.

Another major type of map projection is the *conic projection*, in which the Earth's latitude–longitude grid is transferred onto a cone that is

laid flat and can have one or two standard parallels, depending on how the projection is formed (Fig. 1.2B). Somewhat similar to the conic projection is a *planar projection*, in which an imaginary plane touches the globe at a single point and exhibits a wheel-like symmetry around that point (Fig. 1.3). Planar projections are most commonly used in the representation of polar regions. It is also possible to correctly map shape and area in an interrupted projection, in which the map is not continuous but roughly preserves shape and size (Fig. 1.4).

In addition to containing information on map scale and projection, maps communicate other important information to the cartographer. This information is commonly summarized in a map *legend* (e.g., Fig. 1.5). The map legend provides the meaning of different shapes,

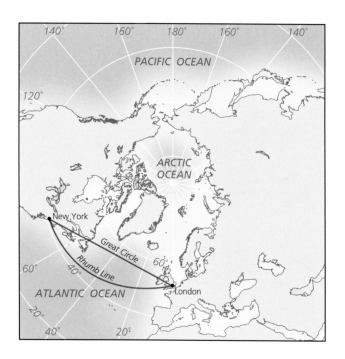

Figure 1.3 The great-circle route from New York to London, which becomes a straight line on this polar gnomonic projection.

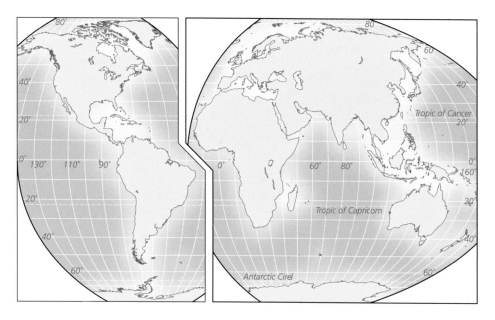

Figure 1.4 Equal-area projection. All mapped areas are represented in their correct relative sizes. This flat polar quartic equal-area projection—in interrupted form—was developed in 1949 for the U.S. Coast and Geodetic Survey by F. W. McBryde and P. D. Thomas.

colors, or shading and helps identify the type of feature being mapped (points, lines, areas, or volumes). Additionally, when rendering volumetric data, it is common to use *isarithmic* *mapping*, which consists of numerous *isolines* (lines of a constant value). Using this process, it is possible to locate areas where the mapped quantity is lower or higher than the values

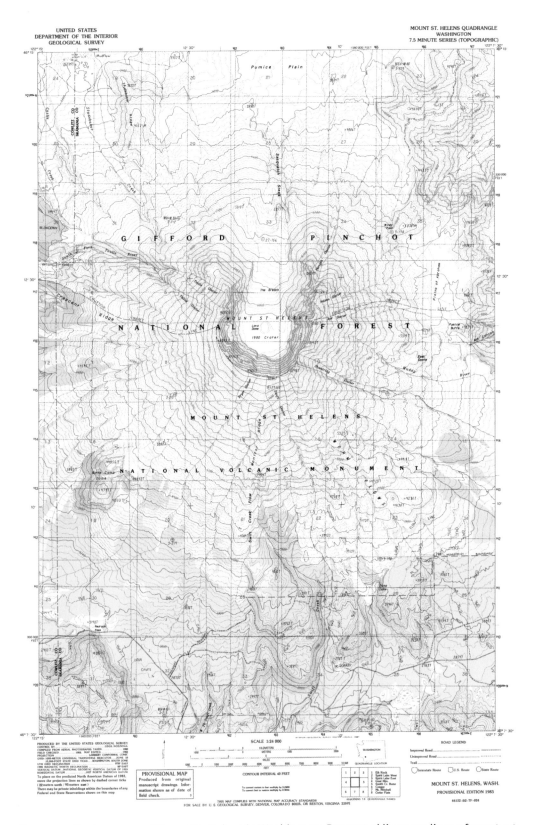

Figure 1.5 Contour map of Mount St. Helens in Washington. Presented lines are lines of constant elevation, thus representing both horizontal position and vertical extent of the terrain. A hi-resolution version of this image is available at www.oup.com/us/mason.

surrounding it. This type of mapping is common for contour lines of topography elevation (e.g., Fig. 1.5) and in meteorological maps (e.g., Fig. 1.6).

Maps contain a significant amount of information about locations on the Earth's surface.

The challenge of the cartographer is to determine the best way to map specific features or information based on the desired outcome. In the exercises below, you will get the opportunity to interact with different types of maps and identify features on those maps.

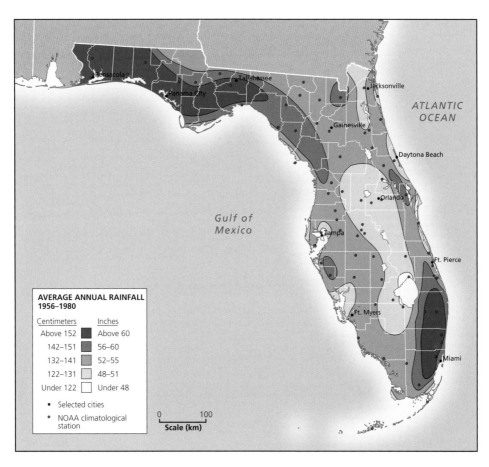

Figure 1.6 Average annual rainfall distribution in Florida, displaying the technique of isarithmic mapping, which allows three-dimensional volumetric data to be shown on a two-dimensional map.

USING THE GLOBE

1) Use your globe to identify the latitude and longitude of the following locations:

    a. Shanghai, China   _____ longitude; _____ latitude

    b. Sydney, Australia   _____ longitude; _____ latitude

    c. Washington, D.C.   _____ longitude; _____ latitude

    d. Moscow, Russia   _____ longitude; _____ latitude

    e. Cairo, Egypt   _____ longitude; _____ latitude

    f. Mexico City, Mexico   _____ longitude; _____ latitude

2) Explain why a map projection is not needed when finding these locations on a globe.

_____

_____

_____

3) Using the scale on your globe, estimate the distance between the following locations:

    a. New York, New York, and Los Angeles, California   _____ miles

    b. Paris, France, and Berlin, Germany   _____ miles

    c. Anchorage, Alaska, and Tokyo, Japan   _____ miles

    d. Bombay, India, and Perth, Australia   _____ miles

THE U.S. ATLAS

4) Identify the type of map projection used in your U.S. atlas. Also, note the scale. Is the atlas a larger scale or smaller scale than the globe?

_____

_____

_____

5) Following the road network on your atlas, determine the distance between the following locations *if you were driving*:

     a. New York, New York, and Hartford, Connecticut   _____ miles

     b. New Orleans, Louisiana, and Dallas, Texas   _____ miles

     c. Phoenix, Arizona, and Salt Lake City, Utah   _____ miles

6) Estimate the driving distance between New York, New York, and Los Angeles, California. How does this compare to the distance you estimated in (3a)? Explain the differences.

_____

_____

_____

TOPOGRAPHIC CONTOUR MAPS

The topographic map below is for the area surrounding the Grand Canyon in Arizona. Using the topographic map for the northern Grand Canyon area (Fig. 1.7), answer the questions that follow.

1) Based on the features you see in the topographic map, what was likely the physical feature responsible for the formation of the Grand Canyon? Explain your reasoning.

_____

_____

_____

2) Where are the lowest points located on this map? Provide a general location. Based on your answer to (1), does this result make sense? Explain your reasoning.

_____

_____

_____

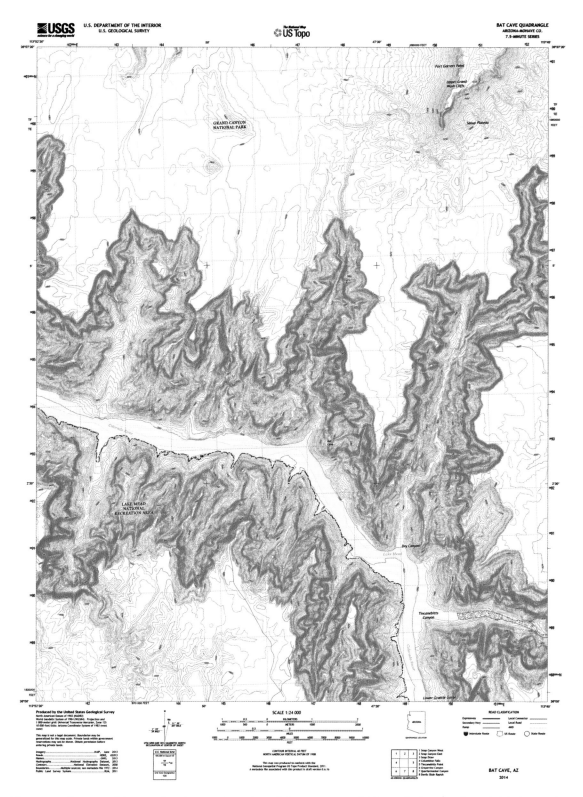

Figure 1.7 Topographic map for the northern Grand Canyon area. A hi-resolution version of this image is available at www.oup.com/us/mason.

3) The *slope* of the terrain (also referred to as the *gradient*) represents how much the elevation of the terrain changes over a given horizontal distance. Using the map scale as an estimate, estimate the slope between Grand Canyon National Park and the Colorado River due south of the location identified on the map for the National Park. Provide this estimate in both feet and miles.

_____ feet _____ miles

4) If you were to walk from Grand Canyon National Park to the Upper Grand Wash Cliffs, would you be walking up or down? How far total would you walk (both horizontally and vertically)? Provide an estimate in feet and in miles.

_____ feet _____ miles

ISARITHMIC MAPS

The map in Figure 1.8 depicts the average daily rainfall accumulation for the United States for December 2011. Answer the following questions about the map.

1) Estimate the average daily amount of rainfall received at each of the locations below:

    a. New Orleans, Louisiana (29.97°N, 90.05°W)   _____

    b. Huntsville, Alabama (34.73°N, 86.58°W)   _____

    c. Little Rock, Arkansas (34.74°N, 92.29°W)   _____

    d. Pensacola, Florida (30.42°N, 87.22°W)   _____

    e. Jackson, Mississippi (32.30°N, 90.18°W)   _____

2) Based on the angle between the latitude and longitude lines in this map, what type of map projection do you expect is being used in this map?

_____

_____

_____

3) Determine how much rainfall, in inches, was received at each of the locations below:

    a. New Orleans, Louisiana (29.97°N, 90.05°W)   _____

    b. Huntsville, Alabama (34.73°N, 86.58°W)   _____

c. Little Rock, Arkansas (34.74°N, 92.29°W) _____

d. Pensacola, Florida (30.42°N, 87.22°W) _____

e. Jackson, Mississippi (32.30°N, 90.18°W) _____

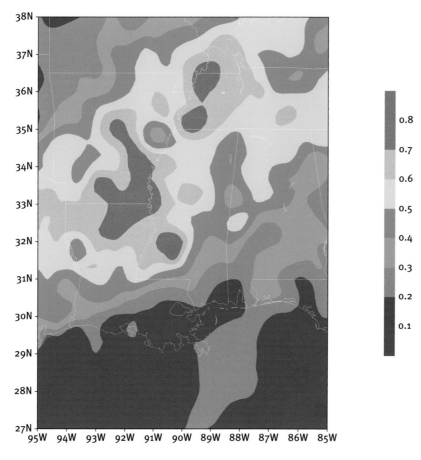

Figure 1.8 Average daily rainfall accumulation (mm) for December 2011.

CRITICAL THINKING QUESTION

1) Many digital GPS systems that are included in smart phones or in vehicles aim to provide as much map information on the screen as possible, without providing many of the
details that are commonly seen in printed maps. For example, map projection information
is nonexistent in most GPS systems used for driving. Do you think map projection information is an important component for a GPS system to have? Why or why not?

_____

_____

_____

GEOGRAPHIC INFORMATION SYSTEMS

GOALS OF THIS LAB:

To understand the advantages of using geographic information systems. To understand a basic geographic information system interface and understand basic map interpretation using a geographic information system.

REQUIRED ITEMS FOR LAB:

A computer with Internet access, a free account with ESRI for ArcGIS. *Note that this lab is designed for use with the ArcGIS online map tool, though the methods should be scalable to other GIS software types.*

A *geographic information system (GIS)* is a computer-based mapping system that is used to overlay multiple layers of information on a map. With the advent of high-power computing technology in the 21st century, computers have been able to handle large quantities of information with relative ease, paving the way for an increased utility of a GIS. GIS software has several direct applications in everyday life, including mapping directions from one location to another, making topographic or meteorological maps, and even identifying locations of people or things with specific characteristics of interest (e.g., mapping the locations of all known clientele for a business). GIS software can map amounts, densities, nearby or distant locations, and quantities and their changes. This lab will provide you with a brief overview of how GIS software works and will give you an opportunity to try some basic GIS analysis.

GIS systems recognize two primary types of data, *vector data* and *raster data*. Vector data describe geographic information with specific

boundaries, such as nations, impact areas, buildings, etc. Vector data fall under three primary subgroups:

- Point data (Fig. 2.1): individual locations on a map (e.g., the position that a lightning strike impacted Earth)
- Poly-line data (Fig. 2.1): vector datasets that have at least two sets of geographic coordinates (latitude–longitude coordinates) that are connected by lines
- Polygon data (Fig. 2.2): vector datasets that have at least three sets of geographic coordinates that are connected and create an enclosed polygon

Vector data can also contain additional information, referred to as *attribute information*. These can be additional details about the datasets, including time of occurrence (in the case of the lightning strike data) or some other descriptive information that represents what the vector is being used to display.

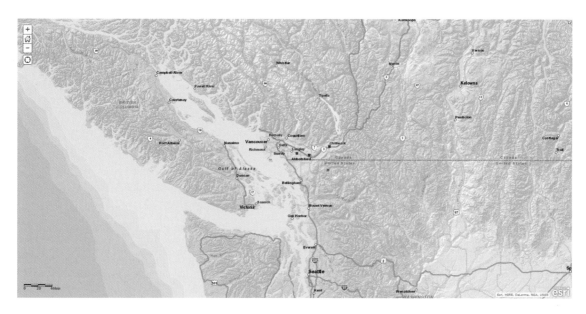

Figure 2.1 Point data for a 2015 avian flu outbreak near Vancouver, British Columbia. Note that the roads presented in Figure 2.1 represent poly-line format vector datasets. A hi-resolution version of this image is available at www.oup.com/us/mason.

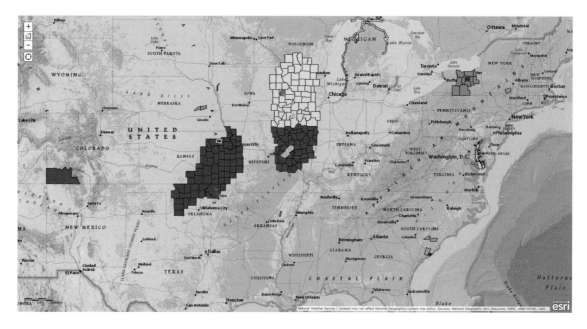

Figure 2.2 Weather watches and warnings valid at 3:30 PM CDT 18 August 2015, presented as polygon GIS vector data. A hi-resolution version of this image is available at www.oup.com/us/mason.

Raster data (e.g., Fig. 2.3) contrast vector data in that they represent continuous spatial fields of information, not discrete points or polygons. Since it is impossible to represent every single point on Earth as a continuous field, data are often translated to a *raster grid*, which is an underlying grid used to create the raster field. Each grid cell in the raster grid is assigned a value based on the value of the continuous or discrete data for that grid cell. Raster grids are often very small to give the visual impression of continuity in the mapped field.

Vector and raster data are stored in unique data format types. The most commonly used data format type for vector data is referred to as a *shapefile*, which retains all geographic and

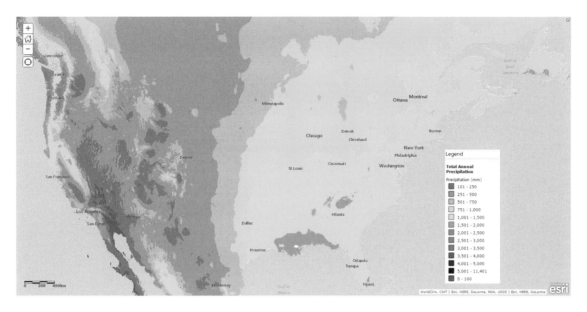

Figure 2.3 Annual average precipitation for the portions of North America from 1950 to 2010. A hi-resolution version of this image is available at www.oup.com/us/mason.

attribute information for the vector dataset. For a raster file, most often the TIFF or JPEG format is used to store the information.

GIS data are often grouped together into collections with at least one similar, dominant attribute that is referred to as the *theme*. When adding datasets with a similar theme, the GIS will add a *layer* of data to the existing *base-map* (original bottom map underlying the GIS data layers), allowing for the new dataset (either raster or vector) to be added to the already existing map with any other layers that may exist. Once integrated into the GIS, layers can be turned on or off at the user's convenience, and even reordered to place one layer above another. It is important to note that some data are *scale dependent*, meaning they are only visible at certain map scales, so even if the layer is turned on, the data associated

with that layer may not be visible without first zooming in. Of course, all GIS map systems have the ability to search for attributes, zoom in or out over specific regions of interest, and move the view around to different regions as needed.

This provides a very brief summary of some of the characteristics associated with GIS software. GIS is an important tool in the 21st century and an invaluable resource to have when doing geographic work. The challenge associated with GIS is not actually operating the software itself, but knowing the types of questions you wish to answer and how to use the GIS software to answer those questions. That challenge can only be met through experience, so use GIS software frequently to master the basics and further understand its significance in geographic science.

THINKING ABOUT GIS DATA

Vector data are discrete data that can be represented as one of three possible subgroups (geometries). Raster data are continuous data. Provide three examples of what the GIS could be used to map for each of these data types.

- Vector Point data

_____

- Vector Poly-line data

_____

- Vector Polygon data

_____

- Raster data

_____

A REAL-WORLD GIS EXAMPLE

We are going to be using the online version of ESRI's ArcGIS software to complete the remaining components of this lab exercise. You need to navigate your computer to www.arcgis.com/home/webmap/viewer.html to access ESRI's online ArcGIS software. In this activity, we will be studying the relationship between soil type, elevation, and flood recurrence interval in the lower Mississippi River Valley region, specifically focusing on the region in eastern Arkansas, western Tennessee, and northwestern Mississippi along the Mississippi River.

First, search for a basemap layer by clicking the Basemap button under My Map and clicking Topographic. Use the mouse to zoom in the map to focus specifically on our region of interest, centered roughly over Tunica, MS, zoomed to a map scale of roughly 15 km per inch of map space. This region is known as the Mississippi Delta region due to the flat terrain swept out by the Mississippi River floodplain. Your map should resemble Figure 2.4.

We will begin this analysis by looking at a layer showing the elevation of our study region. Click the Add button to Search for Layers, then search for elevation. Once identified, add the World Shaded Relief layer to the image. Answer the questions related to this image.

1) What do you notice about the elevation for the region surrounding a large portion of the Mississippi River in the southern half of the domain?

_____

Figure 2.4 Example study domain for this GIS lab assignment. A hi-resolution version of this image is available at www.oup.com/us/mason.

2) One unique feature of the Mississippi River is how curved it is. Often, it becomes so curved that the river cuts off the curve from the main stream and leaves behind an *oxbow lake*, which is simply a lake that used to be part of the Mississippi River. According to this image, how many oxbow lakes have formed from the Mississippi River?

_____

_____

3) Speculate as to the relationship between the prevalence of these oxbow lakes and the elevation map you created.

_____

_____

_____

Let's now add a layer for soil type. Click the Add button and search for agriculture. One layer that comes back is entitled USA Soil Survey. Add this layer to your map. Now, click the Details button under My Map and click the Legend button. Answer these questions:

1) What type of GIS data are represented by this soil dataset? Explain how you know.

_____

_____

2) What is the most dominant soil type east of the Mississippi River in this study region? What about west of the Mississippi River?

_____

_____

3) What soil type seems to be adjacent to nearly every body of water in the map you created?

_____

_____

4) Many of the soil types seem to follow branch-like structures east of the Mississippi River. Many also have lakes tied into them in various locations. Speculate as to the cause of these unique shapes in the soil landscapes of this region.

_____

_____

_____

_____

Finally, we will overlay the 100-year recurrence flood interval data provided by FEMA. A 100-year recurrence flood interval means that floods are expected to occur in these regions once per 100 years, on average. Search for a layer called <u>USDHS FEMA 100 Year Flood Zones</u>. Add this layer and answer the questions that follow.

1) What type of GIS data are these 100-year flood zone data? Be specific.

_____

2) Compare this image with the elevation map you created (toggle back and forth between the two if you wish). What do you notice about the recurrence interval map and the elevation map? Explain this relationship.

_____

_____

_____

3) Similarly, look at the soil type map and the flood recurrence interval map. Do you see any relationships there? Explain.

_____

_____

_____

4) Summarize all of these findings regarding the elevation map, the soil type map, and the flood recurrence interval map in a single paragraph below, discussing specifically which factors you think are most important for describing the 100-year flood recurrence interval. Once finished, you will have completed an interesting geographic study on flood recurrence interval using a GIS.

_____

_____

_____

_____

CRITICAL THINKING QUESTION

Assume for a moment you are interested in determining the number of hurricane landfalls in the states of North and South Carolina from 1990 to 2014. From start to finish, design a simple GIS experiment that will allow you to determine this number, and discuss which dataset(s) you will use. Then, complete the experiment by reading your dataset(s) into the online GIS and determining how many hurricanes made landfall in those two states from 1990 to 2014.

Experimental design:

_____

_____

_____

_____

_____

_____

_____

_____

_____

Results—how many hurricanes made landfall in NC and SC from 1990 to 2014? _____

EARTH–SUN RELATIONSHIPS

GOALS OF THIS LAB:

To understand the basic evolution of the Earth–Sun relationships over the course of a year. To understand how the seasons evolve and the cause for the seasons.

REQUIRED ITEMS FOR LAB:

Textbook and a globe

The Sun is the most important interstellar body for the Earth because its energy supports the different physical processes that occur on our planet. Different Earth–Sun relationships are responsible for the transition between night and day, as well as the transition from season to season. The Earth has two important planetary motions with respect to the Sun, *revolution* and *rotation.*

The Earth's rotation is centered on the Earth's *axis of rotation* that passes directly through the North Pole and the South Pole. The Earth's rotation is counterclockwise, viewed from above the North Pole, meaning it rotates west to east. The Earth completes one entire rotation on its axis in 24 hours. Interestingly, a point on the surface of the Earth moves through space fastest at the equator (nearly 1,100 mph), but at the poles it does not move at all; it merely rotates. At any one time, half of the Earth is illuminated by sunlight while the other half is in darkness. The dividing line forms a great circle known as the *circle of illumination.* Energy from the Sun, known as *insolation* (meaning incoming solar radiation), heats up the land and ocean surface,

which acts to heat the atmosphere and warm the planet.

In addition to the Earth's rotation, the Earth travels around the Sun once every 365 and one-quarter days along its path of orbital revolution. The Earth's orbit is not circular but has an elliptical shape, meaning that the distance between the Earth and the Sun changes as the Earth revolves around the Sun. At *perihelion*, which occurs on January 3 each year, the Earth is nearest the Sun, while at *aphelion*, which occurs on July 4, the Earth is farthest from the Sun. However, the difference between these points is only a fraction of the total Earth–Sun distance, resulting in only a small percentage change in received energy. Every four years, the quarter of a day that appends each orbital revolution is "made up" by adding a day to the year, thereby giving us a *leap year.* However, as with Earth–Sun distance, the number of days in a year does not noticeably contribute to any visible seasonal changes.

The Earth revolves around the Sun along an orbital plane known as the *ecliptic.* The Earth's axis is tilted relative to the *ecliptic* by 23½° and this tilt is constant, meaning that at each point along the ecliptic, the Earth's axis remains parallel to itself (known as *parallelism*). This means that during most of the year, one hemisphere of the Earth tends to point more directly toward the Sun than the other hemisphere. Specifically, during the *summer solstice* (June 22), the Northern Hemisphere is most directly pointed toward the Sun, and the Sun is directly overhead at 23½° N (the *Tropic of Cancer*). Additionally, points north of 66½°N (the *Arctic Circle*) receive 24 hours of continuous daylight while points south of 66½°S (the *Antarctic Circle*) receive 24 hours of continuous darkness.

Roughly six months after the summer solstice, the Sun is directly overhead at 23½°S latitude (the *Tropic of Capricorn*), and so the Southern Hemisphere is receiving more direct sunlight. This is known in the Northern Hemisphere as the *winter solstice* and occurs on or around December 22. Also, during the winter

solstice, points south of the Antarctic Circle receive 24 hours of daylight, while points north of the Arctic Circle receive 24 hours of darkness.

The latitude over which the Sun is directly overhead varies between the Tropic of Cancer and the Tropic of Capricorn, from which it follows that twice a year the Sun is directly overhead at the equator and all latitudes between the two Tropics. On days where the Sun is directly overhead at the equator, there are exactly 12 hours of daylight and 12 hours of darkness for all points on the planet; these days are known as *equinoxes*. In purely astronomical terms, these four transitional dates (two solstices and two equinoxes) mark the beginning of the seasons for each hemisphere and demonstrate the importance of the Earth's axial tilt relative to the ecliptic as the primary cause for our seasons. Figure 3.1 summarizes these important points on the Earth's orbit.

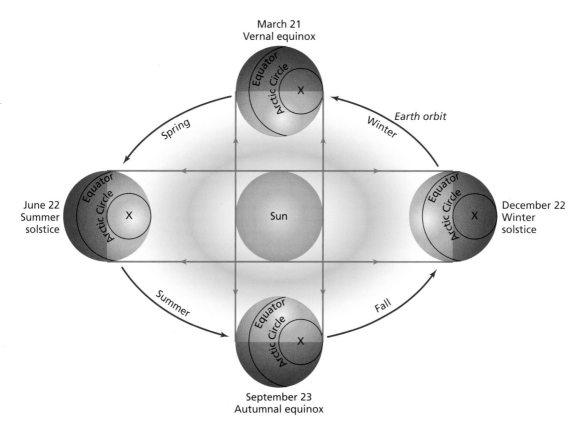

Figure 3.1 March of the seasons as viewed from a position above the solar system. Seasonal terminology here applies to the Northern Hemisphere. (The Southern Hemisphere seasonal cycle is the exact opposite.)

Because the Earth is tilted, different locations have different solar altitudes at various times of the year. For any location, the highest altitude occurs at *solar noon*, when the Sun reaches its highest angle for the day. The latitude at which the Sun is directly overhead will have a noon solar altitude of 90°. Other locations will have a noon solar angle less than 90°. With a lower solar angle, the intensity of sunlight reaching the surface will be lower. As the latitude of the overhead Sun changes seasonally, there is a corresponding seasonal change in solar altitude and incoming radiation and, as a result, a change in length of day. The angle of incidence for the Sun's radiation will affect how warm a location gets. Figure 3.2 shows the relationship between latitude and time of year for daily insolation and helps to explain why certain latitudes on Earth tend to be warmer than average and others cooler at various times of the year.

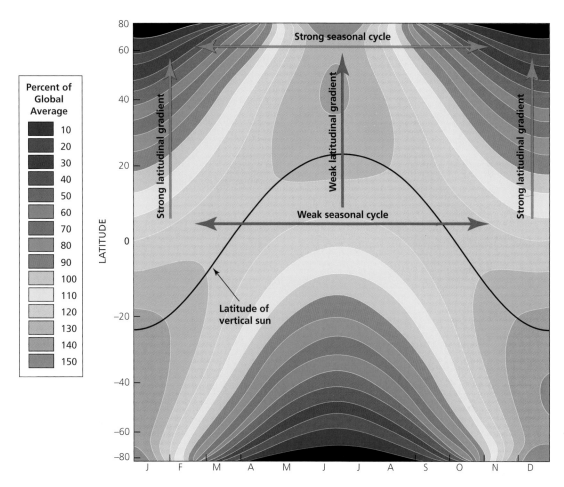

Figure 3.2 Seasonal and spatial variation in solar radiation reaching the top of the atmosphere as a percentage of the global average. The latitude axis is scaled to account for shrinking area at higher latitudes. Compare with patterns in Figure 3.3.

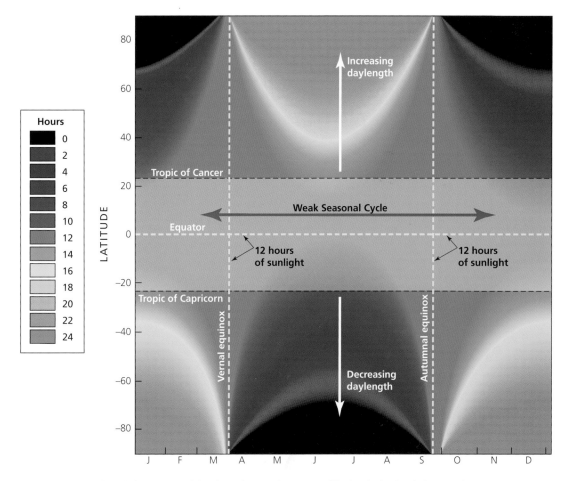

Figure 3.3 Graphical depiction of day length as a function of latitude for both hemispheres.

IDENTIFY THE SEASONS

1) Using your globe, determine what season is occurring at each location for each date provided.

     a. Shanghai, China   July 1   season _____

     b. Sydney, Australia   January 10   season _____

     c. Moscow, Russia   October 21   season _____

     d. Buenos Aires, Argentina   November 1   season _____

     e. Mexico City, Mexico   August 15   season _____

2) Will the Sun ever be directly overhead in Dallas, Texas? Why or why not?

_____

_____

_____

3) State which day(s) of the year the Sun is directly overhead for each of these latitudes.

     a. 0°   _____

     b. 23.5°N   _____

     c. 23.5°S   _____

4) If the Earth rotated on its axis in the opposite direction (east to west), how would the seasons be affected (if at all)?

_____

_____

_____

5) Explain how it is possible that the Earth can be nearest the Sun during the Northern Hemisphere's winter and farthest away during the Northern Hemisphere's summer.

_____

_____

_____

AVERAGE DAILY INSOLATION

Please refer to Figure 3.2.

6) Why do locations such as Hawaii have basically no seasonal changes year to year while places such as Chicago, Illinois, have extreme seasonal changes from year to year?

_____

_____

_____

7) Estimate the average amount of daily insolation received on top of the atmosphere at each of these cities for the month provided as a percentage of the global average.

    a. Trenton, New Jersey (40°N)   February   _____

    b. Memphis, Tennessee (35°N)   April   _____

    c. San Jose, Costa Rica (10°N)   December   _____

    d. Sucre, Bolivia (20°S)   August   _____

    e. Montevideo, Uruguay (35°S)   June   _____

    f. Puerto, Toro, Chile (55°S)   May   _____

8) The dark red sections in Figure 3.4 represent times when there is no insolation received at these locations. Explain why.

_____

_____

_____

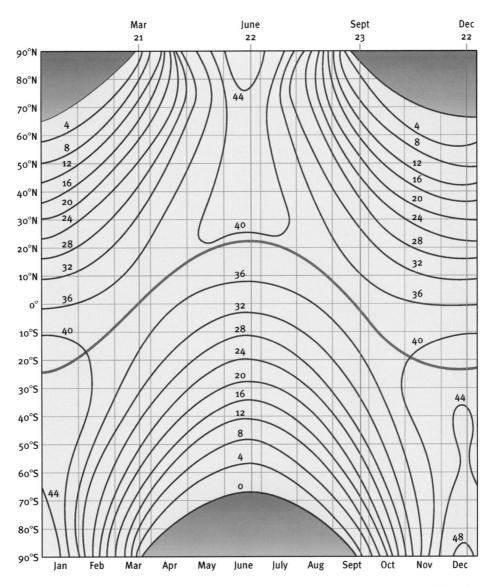

Figure 3.4 Spatial distribution of solar radiation falling on the top of the atmosphere (in megajoules per square meter per day). The annual variations you see are the result of the combined effects of solar elevation and the duration of daily sunlight.

CRITICAL THINKING QUESTION

The Milankovitch cycles represent long-term modifications to Earth–Sun relationships and have been directly linked to long-term natural climate change. There are three Milankovitch cycles:

- Orbital eccentricity: The orbit between the Earth and the Sun modifies from a circular orbit to a more elliptical orbit than we currently have. This modification occurs over the course of roughly 100,000 years.
- Axial precession: The Earth's axis has a precession that causes the North Pole to slowly move in a circular motion, so that the North Pole sometimes points toward Polaris (the North Star) and other times toward Vega. One rotation cycle occurs over roughly 25,000 years.
- Axial tilt: The Earth's axial tilt slowly changes between 22° and 24.5° over the course of about 40,000 years.

1) Given what you now know about the Milankovitch cycles, which of these do you think would have the most profound effect on the Earth's *seasons*? Explain your reasoning.

_____

_____

_____

2) It is known that planets move more slowly when they are farther away from their parent star. Given this information and the fact that the Earth's orbit drastically changes shape every 100,000 years or so, will we be more likely to enter an ice age during a period of high eccentricity or during a period with a circular orbit? Explain your reasoning.

_____

_____

_____

RADIATION AND HEAT BALANCE OF THE SURFACE AND ATMOSPHERE

GOALS OF THIS LAB:

To identify important components of the energy budget for the Earth's atmosphere. To determine how heat is distributed globally and how the atmosphere heats and cools.

REQUIRED ITEMS FOR LAB:

Textbook

The Sun plays a key role in energy distribution on Earth, contributing 99.97 percent of all energy used by physical and biological processes. However, insolation is not distributed evenly over the entire globe, and vast amounts of heat are distributed by atmospheric and oceanic processes to maintain the climates we observe on Earth.

Radiation refers to a net transfer of energy through electromagnetic waves. Electromagnetic radiation is characterized in terms of its energy, which is directly related to the wavelength (distance between wave peaks) or the frequency (how often an entire wave passes a point). Higher-energy radiation has shorter wavelengths and higher frequencies and is referred to as *shortwave* radiation. Lower-energy radiation is *longwave* radiation. The entire spectrum of wavelengths and frequencies is known as the *electromagnetic spectrum* (Fig. 4.1). The radiation emitted from an

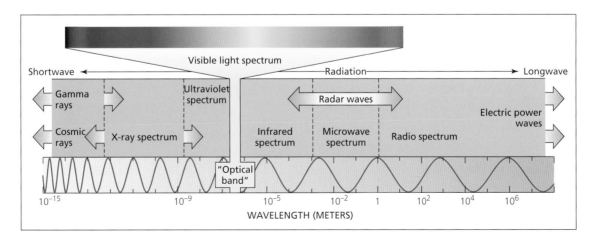

Figure 4.1 The complete (electromagnetic) radiation spectrum. Note the logarithmic axis for wavelength and that wavelengths are not drawn to scale.

object is directly related to its temperature. Cooler objects radiate lower-energy (longwave) radiation (e.g., the Earth), while warmer objects radiate higher-energy (shortwave) radiation (e.g., the Sun). Furthermore, the rate of energy emission over all wavelengths also increases with temperature. For example, a square meter of the Sun emits millions of joules per second (watts), whereas the Earth emits at a rate of only a few hundred watts per square meter. Thus, a hotter object will emit more radiation than a cooler object, and it will emit a higher proportion at shorter wavelengths.

The radiation balance between the Earth and the Sun can be characterized as a summary of six major components:

1. Incoming shortwave radiation from the Sun passes through the atmosphere. A fraction of this radiation reaches the surface and is *absorbed*.
2. Some radiation is absorbed in the atmosphere by clouds, gases, and particulates.
3. The remaining incoming shortwave radiation is *reflected* by clouds or by land and ocean surfaces (especially light-colored surfaces such as snow, sea ice, and sand). This reflection by the land surface is referred to as the Earth's *albedo* and is measured as a *percentage of insolation* that is reflected back into space.
4. The Earth's surface emits low-energy longwave radiation upward. A small part of this radiation passes through the atmosphere and is released by the planet.
5. The vast majority of the longwave radiation emitted by the surface is absorbed by clouds and atmospheric *greenhouse gases*. This absorption increases atmospheric temperature, leading to the observed *greenhouse effect*.
6. The atmosphere emits longwave radiation downward to the surface and upward to outer space. The downward radiation warms the surface. The upward radiation is an energy loss for the planet.

Figure 4.2 summarizes these processes. On a global average basis, roughly 70 percent of incoming solar radiation is absorbed by the planet. For the planet to have an energy balance, an equal amount of longwave radiation is emitted to space by the atmosphere and surface. The surface and atmosphere have very different radiation budgets from one another. The surface absorbs more radiation than it emits, whereas the atmosphere emits more radiation than it absorbs (Table 4.1). Clearly, if they are to have an energy balance, there must be other ways of moving heat between the surface and atmosphere (Fig. 4.3).

When describing the *heat energy balance*, there are four primary components that need to be considered:

1. *Net radiation*—the net accumulated radiation as described by the balances between the longwave and shortwave radiation at a point (e.g., Table 4.1)

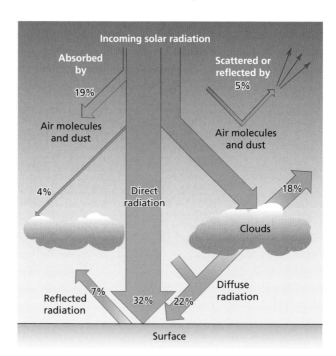

Figure 4.2 Solar radiation flows in the atmosphere.

Radiation Category	Amount of Radiation (%)
Surface	
Solar Absorbed (Gain)	47
Longwave from Atmosphere (Gain)	98
Longwave Emitted (Loss)	−116
Net Radiation	29
Atmosphere	
Solar Absorbed (Gain)	23
Longwave from Surface (Gain)	104
Longwave Emitted (Loss)	−156
Net Radiation	−29
Planet	
Solar Absorbed by Surface (Gain)	47
Solar Absorbed by Atmosphere (Gain)	23
Longwave from Atmosphere (Loss)	−58
Longwave from Surface (Loss)	−12
Net Radiation	0

TABLE 4.1 Global Average Radiation Amounts as a Percentage of Incoming Solar Radiation

2. *Sensible heat*—transferred from surface to atmosphere in a form that can be sensed—that is, by raising the temperature of the atmosphere

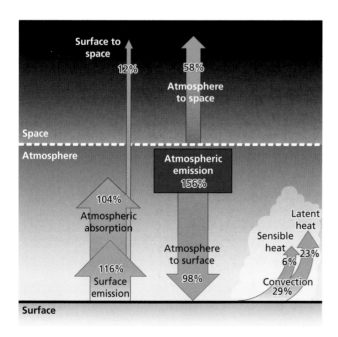

Figure 4.3 Global mean terrestrial radiation and convection. All values are expressed as a percentage of solar radiation entering the top of the atmosphere.

3. *Latent heat*—heat transferred from surface to atmosphere when water evaporates from the surface. In general, latent heat refers to energy required to do a phase change, but the most common phase change near the Earth's surface is the water–vapor change (evaporation) or the vapor–water change (condensation).

4. *Ground heat*—the transfer of heat from the surface downward into the Earth's soil or other surface objects (buildings, oceans, trees, etc.)

The net radiation accumulated at a point is distributed via sensible heating, latent heating, and ground heating, to ensure that any location on Earth maintains an energy balance. However, the climate of different locations dictates how this heat distribution occurs at each point. Dry locations will tend to have less latent heating, since little water is available for evaporation. This means that these locations must achieve balance through

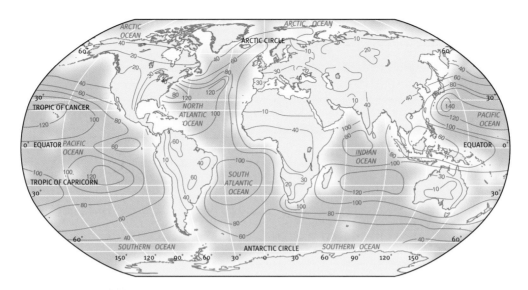

Figure 4.4 Global distribution of latent heat loss. The heat used in evaporation is expressed in thousands of calories per square centimeter per year. Red isolines show the pattern over land, and blue isolines show the pattern over the oceans.

increased sensible heating. Globally, roughly 70 percent of the net radiation accumulated at the surface is distributed through latent heating, while 30 percent is distributed through sensible heating. Figure 4.4 summarizes the net sensible and latent heat values annually for the entire globe. It is important to remember that for the most part the Sun does not directly heat the atmosphere. Instead, the Sun heats the surface, and the surface warms the atmosphere by emission of longwave radiation and by sensible and latent heat.

GLOBAL RADIATION BUDGET

1) Based on the annual distribution of radiation at the Earth's surface, as shown in Figure 4.5, which locations on Earth should be the hottest? Which locations should be the coldest? Why?

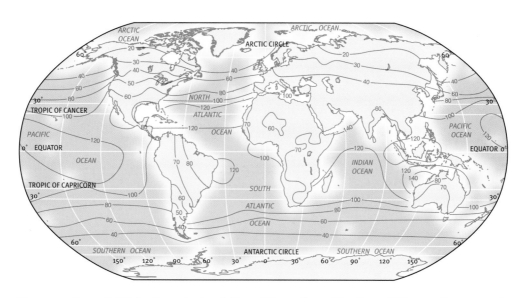

Figure 4.5 Annual distribution of net radiation at the surface of the Earth. Values are in thousands of calories per square centimeter. Red isolines show the pattern over land, and blue isolines show the pattern over the oceans.

_____

_____

_____

2) The theory of *global warming* states that an increase of the greenhouse gas concentrations of the atmosphere would cause a net warming of the atmosphere globally. Explain how increasing greenhouse gases lead to warming.

_____

_____

_____

3) Figure 4.6 shows the global albedo for March 1, 1987. Answer the questions below based on this image.

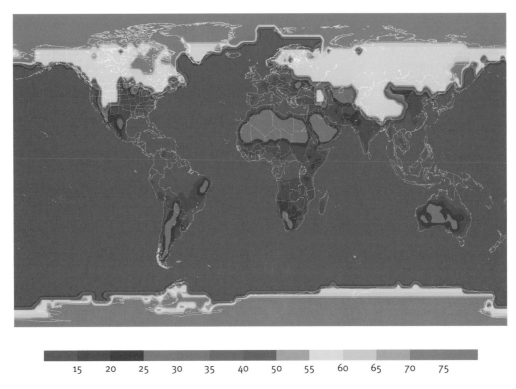

Figure 4.6 Surface albedo from March 1, 1987. Shaded colors are percentages of radiation reflected by the surface.

a. Why is the albedo so low (less than 15 percent) over the oceans?

_____

_____

_____

b. Explain the extremely high percentages over Antarctica and the Arctic Ocean.

_____

_____

_____

c. Explain the locally high albedo feature over northern Africa. A similar feature occurs in Australia and portions of southern South America. What land feature could be responsible for these patterns in the albedo?

_____

_____

_____

d. A significant percentage of Russia shows an extremely high albedo. Similarly, portions of western China and northern India have high albedos, as does most of Canada (on this day). What could be some of the reasons for these high albedo values in these locations?

_____

_____

_____

SURFACE ENERGY BUDGET

4) Using the images in Figures 4.4 and 4.7, estimate the annual heat loss due to latent and sensible heating for the following locations:

    a. New Orleans, Louisiana   _____ sensible; _____ latent

    b. Phoenix, Arizona   _____ sensible; _____ latent

    c. New York, New York   _____ sensible; _____ latent

    d. Miami, Florida   _____ sensible; _____ latent

    e. Detroit, Michigan   _____ sensible; _____ latent

    f. Kansas City, Missouri   _____ sensible; _____ latent

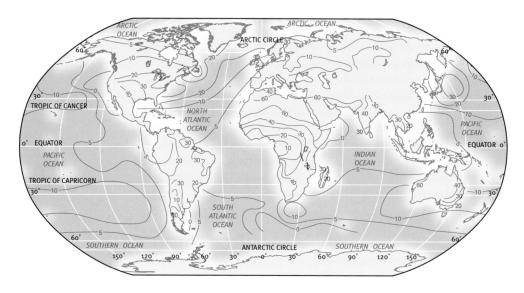

Figure 4.7 Global distribution of sensible heat loss. Values are in thousands of calories per square centimeter per year. Red isolines show the pattern over land, and blue isolines show the pattern over the oceans.

5) Which of the locations in (4) seem to have dry climates? Explain how you can determine this based on your estimates in (4).

_____

_____

_____

6) Table 4.2 lists the monthly net radiation, sensible heating, and latent heating (MJ/m²/ day) for two *hypothetical* cities. Plot these values on the two blank charts (Figs. 4.8 and 4.9). Describe the types of climates suggested by each city.

_____

_____

_____

TABLE 4.2

Month	City 1			City 2		
	Net radiation	Sensible	Latent	Net radiation	Sensible	Latent
Jan	−2	−1	0	8	7	1
Feb	1	0	1	6	6	0
Mar	3	0	2	6	5	1
Apr	4	1	3	5	4	0
May	5	1	4	3	3	0
Jun	6	2	4	2	2	0
Jul	6	1	5	1	1	-1
Aug	4	1	3	3	2	0
Sep	3	1	2	5	3	1
Oct	2	0	2	6	5	2
Nov	0	−1	1	8	6	2
Dec	−1	−2	0	9	6	2

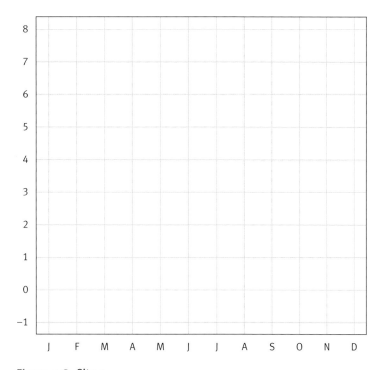

Figure 4.8 City 1.

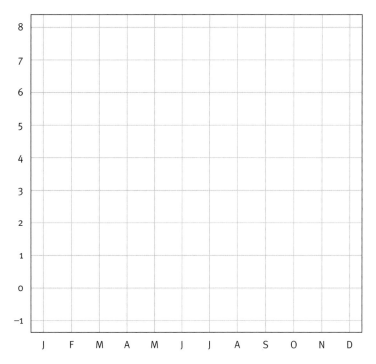

Figure 4.9 **City 2.**

CRITICAL THINKING QUESTIONS

1) The planet Venus has thick clouds of carbon dioxide that cover the entire planet, as well as a runaway greenhouse effect that allows basically no longwave energy to leave the planet. Based on this information, contrast the radiation budget on Venus with that on Earth.

_____

_____

_____

2) Consider the surface energy budget heat characteristics associated with an environment with dry, desert air blowing over an oasis. Discuss the latent and sensible heat characteristics of this environment.

_____

_____

_____

ATMOSPHERIC AND SURFACE TEMPERATURE

GOALS OF THIS LAB:

To understand horizontal and vertical distributions of temperature and the concepts of atmospheric stability and instability. To understand annual and diurnal variations in temperature and heating and the relationships of temperature with insolation.

REQUIRED ITEMS FOR LAB:

Textbook, atlas of the United States, a globe, and a computer with Internet access (for hi-resolution images)

The air around you consists of billions of molecules that are moving in random, chaotic motions. The energy associated with these motions, or any motion, is called *kinetic energy*. Faster motions are associated with higher kinetic energy, and the way we measure the kinetic energy of the motion of molecules is by measuring *temperature*. Thus, temperature is a somewhat abstract idea that measures the energy in a substance.

Because we cannot observe molecular motions easily, we often use substances whose thermal properties we know and observe changes in these substances as a guide to measuring temperature. We call these instruments *thermometers*. Thermometers are calibrated to measure temperature on one of three well-known temperature scales: Fahrenheit, Celsius, and Kelvin. It is fairly easy to convert between the different temperature scales by using these formulas:

There are important temperatures on each scale as well. The boiling point of water occurs at 373 K, 212°F, and 100°C. The freezing point of water occurs at 273 K, 32°F, and 0°C. Additionally, 0 K represents absolute zero, or the temperature at which all kinetic energy of the molecules becomes zero. This temperature does not occur on Earth without the assistance of a laboratory.

In the lowest layer of the atmosphere, the *troposphere*, temperature typically decreases with height. On average, this change in temperature with height, known as a *lapse rate*, is −6.5°C/km. While this is the standard lapse rate, layers within the troposphere often exhibit different lapse rates at any given time, and differences in the lapse rates can lead to changes in *atmospheric stability*. When assessing the stability of the troposphere, we use what is known as *parcel theory*, where we determine how a small parcel (or blob) of air will change in a troposphere with a lapse rate that is

1. Fahrenheit to Celsius   T(Celsius) = (5/9) * (T(Fahrenheit) − 32)

2. Celsius to Fahrenheit   T(Fahrenheit) = (9/5) * T(Celsius) + 32

3. Celsius to Kelvin   T(Kelvin) = T(Celsius) + 273

observed. We make one big assumption in this assessment: We assume the parcel does not gain or lose heat from the surrounding environment, meaning we assume that we have an *adiabatic* process.

When considering how the temperature of an air parcel changes in the troposphere as that parcel is lifted or lowered, we have to think about how temperature and pressure are related. Because air pressure *always decreases with height throughout the atmosphere*, the air temperature of a parcel will decrease as the parcel expands in the lower-pressure air. If the air parcel is not saturated with water vapor, we know that it will cool at a constant rate of 10°C/km, known as the *dry adiabatic lapse rate (DALR)*. However, if the air parcel *is* saturated with water vapor, we know that it cools at a slower, nonconstant rate, known as the *saturated adiabatic lapse rate (SALR)*. We typically assume that the SALR is –6°C/km, though the actual rate is slower than 6°C/km near the surface and nearly the same as the DALR once you approach the *tropopause*, the top of the troposphere.

To assess stability, we must compare the lapse rate of a layer of the troposphere, known as the *environmental lapse rate (ELR)*, to the parcel lapse rate (either the DALR or SALR, depending on if the parcel is saturated). We can lift a parcel from the surface to some height and figure out its temperature based on the DALR or SALR, then compare the parcel temperature to the environmental temperature. If the parcel is warmer, the parcel will continue to rise (because it is less dense), meaning the air is unstable. If the parcel is cooler, the parcel will sink, meaning the air is stable. If the parcel is cooler than the environment when it is unsaturated but warmer when it is saturated, the environment is known as conditionally unstable, and a saturated parcel will rise while an unsaturated parcel will sink.

In addition to these vertical changes, temperature varies horizontally due to *diurnal cycles* (e.g., sunrise vs. sunset and daytime vs. nighttime) and *annual cycles* (e.g., seasons and changes in the highest solar angle as the Earth revolves around the Sun). Additionally, the amount of energy required to increase the temperature of water is considerably more than that of the land, so areas with a nearby body of water (*maritime effects*) tend to have cooler overall temperatures in summer than places that are surrounded by land on all sides (*continental effects*). In winter the ocean cools less than the land at the same latitude and is therefore usually warmer. We can observe the impacts of

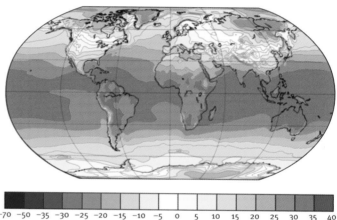

-70 -50 -35 -30 -25 -20 -15 -10 -5 0 5 10 15 20 25 30 35 40
A January surface air temperature

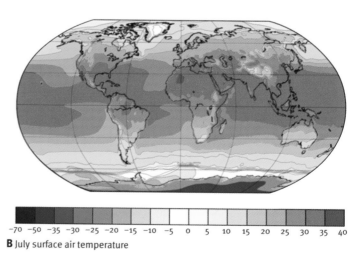

-70 -50 -35 -30 -25 -20 -15 -10 -5 0 5 10 15 20 25 30 35 40
B July surface air temperature

Figure 5.1 Average air temperatures in degrees Celsius for January (A) and July (B).

continental and maritime effects in combination with the annual cycles by using *climographs*. Climographs show how temperature varies monthly for a particular location, and it is easy to diagnose proximity to water by observing small variations in annual temperature (relative to a continental location).

Figure 5.1 shows the global average temperature for January and July, characterized by lines of constant temperature (*isotherms*). The pattern of isotherms matches very closely to the pattern of insolation seen on Earth. Locations that are receiving more direct sunlight are warmer (as expected). These images also show where temperature changes rapidly with latitude; this change of horizontal temperature is known as a *temperature gradient*. When winds blow over these temperature gradients, they move air of one temperature into an area with air of a different temperature. We call this horizontal transport of temperature by winds *temperature advection*. Clearly, the processes involved in the distribution of horizontal and vertical temperature are complex, but they are essential to understanding basic atmospheric processes.

THERMOMETERS AND THE TEMPERATURE SCALES

1) Use your thermometer to measure the air temperature inside of your classroom in Fahrenheit. Note that temperature below. Then, use the formulas in the lab to convert that temperature to Celsius and Kelvin. Is the temperature in your classroom near absolute zero?

_____ Fahrenheit _____ Celsius _____ Kelvin

2) Do the same measurement outside. Is the kinetic energy of the air molecules higher outside or inside? Explain how you know.

_____ Fahrenheit _____ Celsius _____ Kelvin

VERTICAL TEMPERATURE DISTRIBUTIONS

3) Using the three blank vertical temperature profiles in Figures 5.2, 5.3, and 5.4, determine if each given environmental lapse rate is stable, unstable, or conditionally unstable. Explain how you know.

    a. ELR of 7°C/km

	T (environment)	T (parcel,dry)	T (parcel,moist)
2 km	_____	_____	_____
1.5 km	_____	_____	_____
1 km	_____	_____	_____
0.5 km	_____	_____	_____
0 km	20° C	20° C	20° C

Figure 5.2 Blank vertical temperature profile.

b. ELR of 12°C/km

	T (environment)	T (parcel,dry)	T (parcel,moist)
2 km	_____	_____	_____
1.5 km	_____	_____	_____
1 km	_____	_____	_____
0.5 km	_____	_____	_____
0 km	20° C _____	20° C _____	20° C _____

Figure 5.3 Blank vertical temperature profile.

c. ELR of 4°C/km

	T (environment)	T (parcel,dry)	T (parcel,moist)
2 km	_____	_____	_____
1.5 km	_____	_____	_____
1 km	_____	_____	_____
0.5 km	_____	_____	_____
0 km	20° C _____	20° C _____	20° C _____

Figure 5.4 Blank vertical temperature profile.

d. Identify which is stable, unstable, or conditionally unstable, and why:

_____

_____

_____

4) Identify the temperature inversion on the graph in Figure 5.5 (temperature increasing with height in the troposphere). Is a temperature inversion stable, unstable, or conditionally unstable? How do you know?

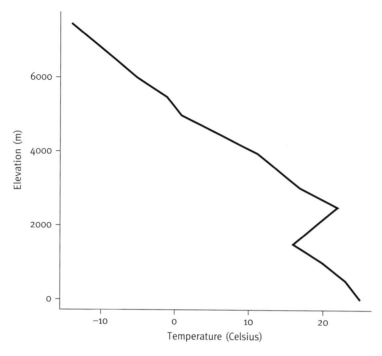

Figure 5.5 Temperature versus elevation.

_____

_____

_____

HORIZONTAL TEMPERATURE DISTRIBUTION

5) Using Figure 5.1, identify regions where the maximum average temperature difference between July and January is greatest and least. Explain what the reasons for these changes could be.

_____

_____

_____

6) For the four climographs in Figure 5.6 (from NCDC data), match them to their location. Explain why you chose each location.

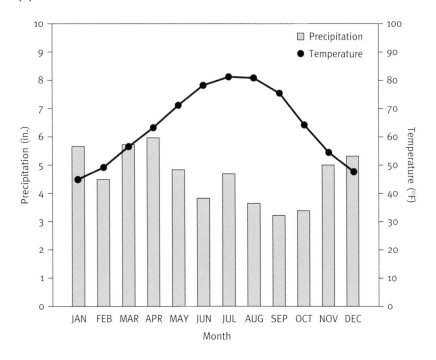

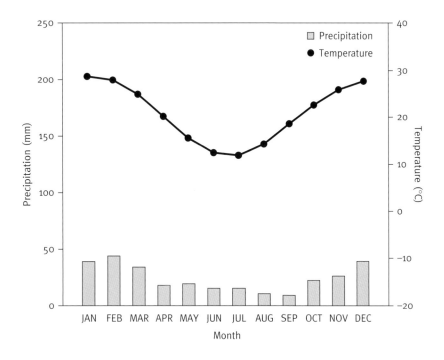

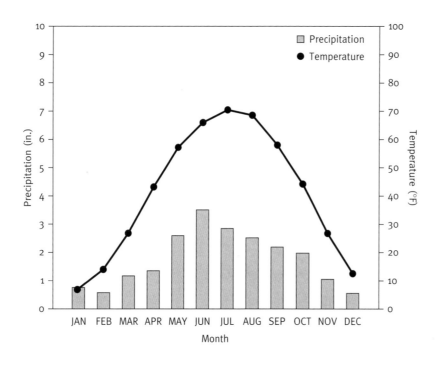

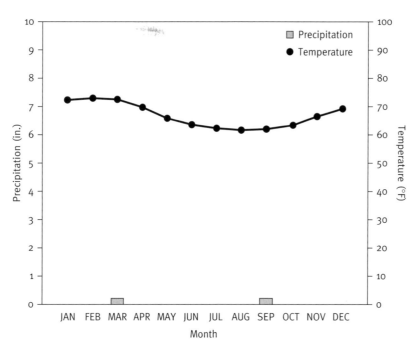

Figure 5.6 Monthly temperatures and precipitation.

i. Lima, Peru _____

ii. Fargo, North Dakota _____

iii. Jackson, Mississippi _____

iv. Outback Desert, Australia _____

Explanation:

_____

_____

_____

_____

_____

_____

_____

_____

CRITICAL THINKING QUESTIONS

The image in Figure 5.7 represents a Skew-T/Log-P chart, which is the chart that is typically used by weather forecasters to depict the vertical temperature profile obtained from a weather balloon launch. In this image, there are several types of lines (which are labeled). The DALR lines (dashed red lines) represent the dry adiabatic lapse rate at the given temperature and pressure, while the SALR (dot-dashed green lines) represent the saturated adiabatic lapse rate at the given temperature and pressure. The thick solid red line represents temperature (°C) while the solid green line represents the dewpoint (°C), which is the temperature at which the air is saturated. Use this Skew-T chart to answer the questions below.

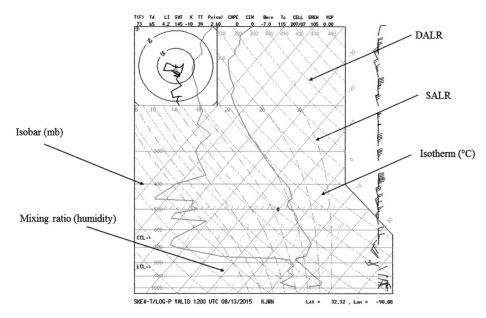

Figure 5.7 Skew-T/Log-P chart for Jackson, MS, on August 13, 2015, at 1200 UTC. A hi-resolution version of this image is available at www.oup.com/us/mason.

1) Are any layers in this atmospheric profile saturated? Explain your reasoning.

_____

2) There are two inversions on this profile below the 500 mb pressure level. Identify those inversions by circling them.

3) Identify a layer of the atmosphere that is absolutely unstable, conditionally unstable, and absolutely stable. Explain your reasoning for the layers you selected. Look below the 500 mb level.

    Absolutely stable layer: _____ mb to _____ mb

    Absolutely unstable layer: _____ mb to _____ mb

    Conditionally unstable layer: _____ mb to _____ mb

Explanation:

_____

_____

4) In the *stratosphere*, the layer of atmosphere above the troposphere, temperature increases with height. The level that marks the division between the stratosphere and troposphere is the *tropopause*. Estimate the level of the tropopause based on this Skew-T.

_____

CIRCULATION PATTERNS OF THE ATMOSPHERE

GOALS OF THIS LAB:

To learn the basic patterns of the atmospheric circulation and their role in local and regional climate.

REQUIRED ITEMS FOR LAB:

Textbook, globe

Understanding global circulation is best attained through the use of models of the atmospheric circulation. When including Earth's rotation, the global circulation is characterized by a *three-cell model* as depicted in Figure 6.1. The global circulation can be outlined in a series of seven belts/important latitudes:

1. Intertropical convergence zone (ITCZ)—belt of low pressure located near the equator that moves seasonally and brings heavy rainfall to regions under its influence
2. Trade winds—steady, easterly winds that generally exist in the tropical latitudes of each hemisphere
3. Subtropical high—occurs at roughly 30°N/S latitude, a semipermanent location of high pressure often corresponding to the world's deserts

4. Midlatitude westerlies—belt of westerly winds in the midlatitudes (roughly 30°N–60°N) that is influenced by both tropical air masses and polar air masses (most likely to experience seasons)
5. Polar front/jet—boundary between the midlatitude westerlies and the polar easterlies that corresponds to a significant exchange of relatively warm midlatitude air and cold polar air. Associated with the strongest of the jet streams in both hemispheres.
6. Polar easterlies—easterly wind belt that flows around each pole and each polar high
7. Polar high—dome of high pressure over each pole, a thermally driven high-pressure center

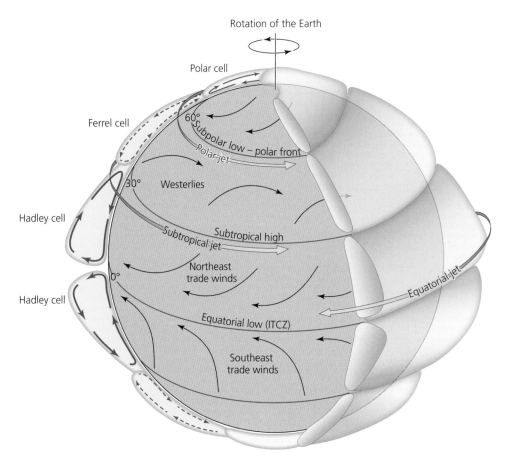

Figure 6.1 A conceptual model of the global atmospheric circulation pattern showing the major surface pressure belts, the prevailing surface wind systems that develop from them, the upper-level jet streams, and the Coriolis deflection of surface winds.

Each of these components shifts seasonally due to the tilt of the Earth. The ITCZ is often associated with the latitude receiving the most direct solar radiation, which shifts seasonally from the Tropic of Cancer/Capricorn in the June/December solstice and the equator during the equinoxes. As this occurs, the remaining components of the circulation shift similarly. These changes can be seen by analyzing isobaric maps each month as outlined in Figure 6.2.

In addition to these horizontal flow patterns, vertical patterns exist in the flow between belts that contribute to the transfer of heat from the tropics to the poles. One circulation that is thermally direct—that is, it moves warm air upward and cold air downward—is known as a *Hadley cell* and occurs between the ITCZ and the subtropical high in both hemispheres (Fig. 6.3). Similarly, a circulation between the polar front/jet and the polar high in both hemispheres is referred to as the *polar cell*. Circulations that are thermally indirect, which move relatively warm air downward and cool air upward, affect the midlatitudes and are called *Ferrel cells*. These cells are important for heat transfer and global energy balance.

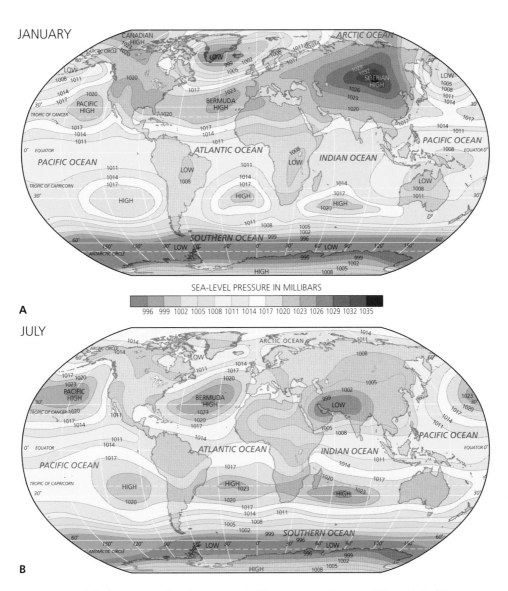

JANUARY

SEA-LEVEL PRESSURE IN MILLIBARS

996 999 1002 1005 1008 1011 1014 1017 1020 1023 1026 1029 1032 1035

A

JULY

B

Figure 6.2 Global mean sea level pressure (mb) patterns in January (A) and July (B).

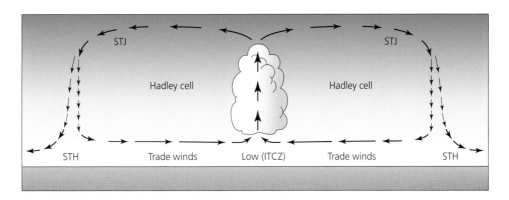

Figure 6.3 A cross-section of the Hadley cell circulation. Surface convergence of the trade winds feeds convective lifting in updraft zones that are embedded along the equatorial low (ITCZ). Westerly acceleration of upper-level winds leads to the formation of the subtropical jet stream (STJ), which initiates convergence of air and general subsidence at about 30° latitude. The resulting subtropical high (STH) is a broad and continuous belt of high pressure that encircles Earth.

THE GENERAL CIRCULATION

1) Using Figure 6.4, fill in the wind flow arrows that correspond to how the winds move in the global circulation (Fig. 6.1 may help with this). Then, match the name of each belt/region with its corresponding name below.

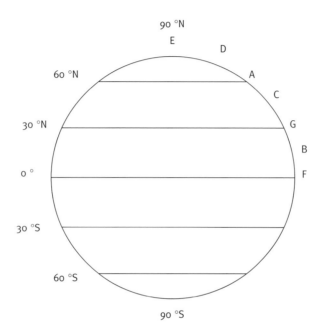

Figure 6.4 Blank global circulation belt map.

_____ i) Subtropical high

_____ ii) ITCZ

_____ iii) Polar high

_____ iv) Midlatitude westerlies

_____ v) Polar easterlies

_____ vi) Polar front

_____ vii) Trade winds

2) Why is the pressure so abnormally above average in the north Atlantic in the summer (July) but not in the winter (January), as seen in Figure 6.3? Guess as to the importance of the size of this area of high pressure (known as the Bermuda high) in hurricane formation and tracking.

_____

_____

_____

3) It is possible, though a bit challenging, to identify the occurrence of certain features in the general circulation through the use of climographs. Using the climographs in Figures 6.5, 6.6, and 6.7, describe the changes in seasonal climate for the locations provided below. Use your globe to identify the locations of these cities. Mention any changes in influences of the different global circulations in your descriptions.

   a. Trivandrum, India

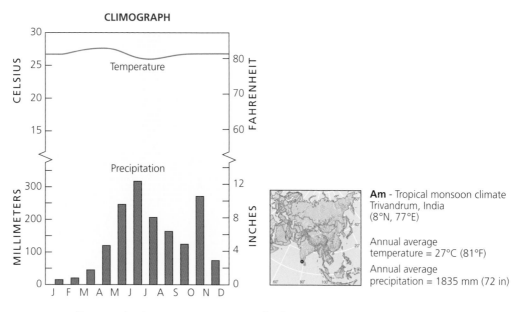

Figure 6.5 Climograph of a monsoon rainforest (Am) weather station.

_____

_____

_____

b. Nikhata Bay, Malawi

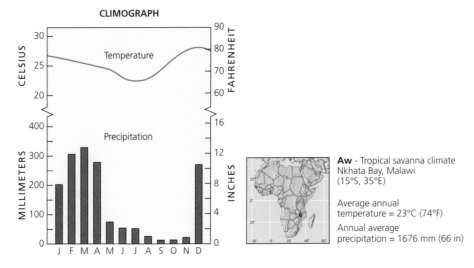

Figure 6.6 Climograph of a tropical savanna (Aw) weather station.

_____

_____

_____

c. Astana, Kazakhstan

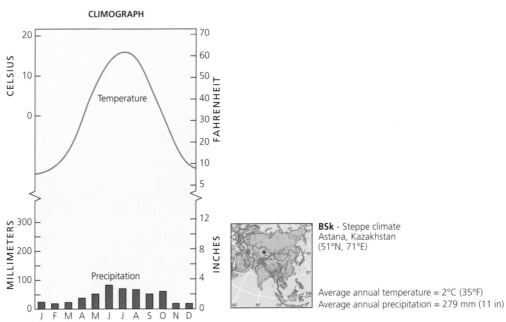

Figure 6.7 Climograph of a midlatitude steppe (BSk) weather station.

_____

_____

_____

4) People have often characterized the Arctic as a desert because it receives even less precipitation in a year than places that are traditionally thought of as deserts (e.g., the Sahara Desert). Explain, using the three-cell general circulation model, why so little precipitation occurs over the Arctic, even relative to a desert.

_____

_____

_____

5) The *meridional* (north–south) flow changes in the polar jet often result in the intrusion and interaction of cold polar air and warm subtropical air. These boundaries are called fronts and often occur collocated with extratropical cyclones. Explain how these changes in the polar jet stream cause the midlatitudes to experience seasons.

_____

_____

_____

CRITICAL THINKING QUESTIONS

The Coriolis effect is responsible for the formation of the three-cell circulation that we observe in each hemisphere. It is an apparent effect that causes air to be deflected in the direction opposite the direction of rotation of the Earth. We are going to use the globe to visualize the Coriolis effect.

1) Look at the globe from the North Pole downward. Begin spinning the globe in a counterclockwise direction. Continue spinning the globe in the same direction, but now flip over the globe and look at it from the South Pole downward. What direction is it rotating?

_____

2) Does this result suggest that air is deflected in the same direction in both hemispheres? If not, which direction do you think the air is deflected in each hemisphere? Explain.

_____

_____

_____

3) Looking at the globe spinning again, at which latitude(s) do you think the Coriolis effect is strongest? Why?

_____

_____

4) Using Figure 6.8, redraw the circulation from problem 1 in the General Circulation section for an Earth that rotates in the opposite direction.

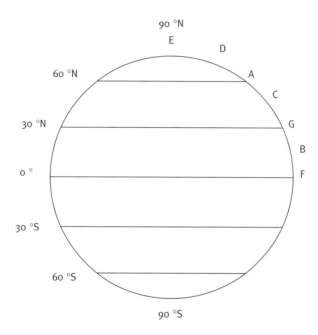

Figure 6.8 Blank global circulation belt map.

ATMOSPHERIC MOISTURE AND THE WATER BALANCE

GOALS OF THIS LAB:
To understand the basic quantities associated with phase changes of water and their impact on the atmosphere. To consider the components of the water budget and quantify their impacts.

REQUIRED ITEMS FOR LAB:
Textbook, calculator, hot water, three ice cubes, a thermometer, an empty clear glass, and a dry erase marker

Water is a substance that is uniquely prevalent on Earth and is required for a multitude of physical and biological processes. Water has a sticky quality due to the hydrogen bonds and the polarity of water molecules associated with it. However, water has some other unusual qualities as well. For example:

1. The solid form is less dense than the liquid form and thus floats in the liquid form.
2. It exists in a liquid form at most surface temperatures and pressures on Earth.
3. The stickiness of water gives it a strong surface tension, as well as capillarity (water can actually defy gravity and move upward in plant roots and stems).

While water primarily exists in liquid form on Earth, it undergoes phase changes as it passes through the *hydrologic cycle*. These phase changes require energy, known as *latent heat*, to break the hydrogen bonds and allow the phase changes to occur. The amount of latent heat required for each phase change is outlined in Figure 7.1.

When water exists in vapor form, we call it *water vapor*. Water vapor is measured in the Earth's atmosphere through one of several measures of *humidity*. For example, water vapor is often quantified by its contribution to overall air pressure, or *vapor pressure*. As air becomes warmer, more water vapor is required to reach a balance between evaporation and condensation. When that limit has been reached, the air is said to be *saturated*, and the corresponding vapor pressure is called the *saturation vapor pressure*. Figure 7.2 shows the relationship between *saturation vapor pressure* and temperature.

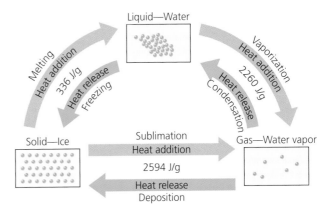

Figure 7.1 The amount of heat added or removed to complete each of the six possible phase changes of water.

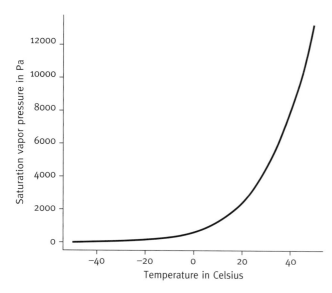

Figure 7.2 The relationship between saturation vapor pressure and temperature.

One of our most common measures of humidity, *relative humidity*, is the ratio of vapor pressure to saturation vapor pressure. When the two are equal, the air is saturated, and water droplets form in the air mass through condensation. Eventually, assuming that sufficient available cloud droplets exist, rain or other forms of precipitation will result, all of which are important components of the hydrologic cycle (Fig. 7.3).

The other important component of the hydrologic cycle is *evaporation*. Areas where significant evaporation occurs without compensating precipitation have deficits in surface water, meaning they experience very dry conditions (drought). Assuming water is available at the surface, evaporation is enhanced through warmer temperatures (which have more energy for latent heat of evaporation) and extremely dry air conditions (so that a gradient in water vapor exists that tries to balance through evaporation of surface water). Other sources of surface water include water passing through the leaves of plants when they undergo respiration. Evaporation of this type of surface water is referred to as *evapotranspiration*. In spite of the consistent evaporation that occurs across all land surfaces, precipitation rates are often fast enough that the water that falls will move across the land surface into a drainage stream, such as a river. The water that is not evaporated or absorbed into the groundwater system is referred to as *runoff*.

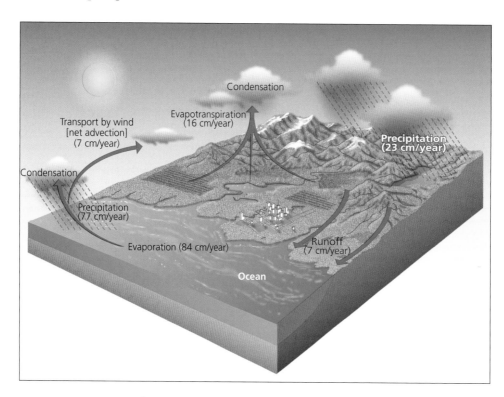

Figure 7.3 A graphical depiction of the hydrologic cycle.

LATENT HEAT

1) In Figure 7.1, the amount of heat required to change between the different phases of water is measured in joules. By definition, specific heat is the energy required to warm 1 kg of a substance (e.g., water) by a temperature of 1°C. However, to change the phase of 1 kg of water requires considerably more heat than changing the temperature, as is seen in Figure 7.1. Using this figure, determine how much energy is required to do the following phase changes:

    a. 5 g of liquid water to ice: —————— J

    b. 15 g of ice to water vapor: —————— J

    c. 15 g of ice to liquid water and then from liquid water to vapor   —————— J

    d. 1 kg of vapor to liquid water   —————— J

2) As a rule of thumb, phase change processes that cause the molecules within a substance to move from a lower-energy state (e.g., solid) to a higher-energy state (e.g., vapor) require that energy be *added* to those systems. These are referred to as *endothermic*. The opposite is true when going from a fast molecular motion to a slower motion (e.g., gas to solid— *exothermic*). Based on these ideas, explain how evaporation is actually a cooling process for the atmosphere. Which other phase changes would be cooling processes?

————————————————————————————————

————————————————————————————————

————————————————————————————————

MEASURES OF HUMIDITY

3) The most commonly used measure of humidity is relative humidity, which measures the percent proximity to saturation of an air parcel. The relative humidity is calculated by:

$$RH = \frac{\text{vapor pressure}}{\text{saturation vapor pressure}} \times 100\%$$

Another measure of moisture is *dewpoint temperature*, which is simply the temperature at which saturation occurs. In other words, the dewpoint is the temperature to which the air must be cooled to cause saturation. To put it differently, Figure 7.1 shows the relationship between vapor pressure and dewpoint. One can use a curve such as the one provided to determine the relative humidity from a temperature (T) and a dewpoint (T_d). To do so, first find the saturation vapor pressure for the air for a given temperature. Next, find the vapor pressure based on the dewpoint using the same curve. The ratio of these vapor pressures is the relative humidity. Using the curve provided in the lab, estimate the relative humidity for the following temperature/dewpoint combinations:

    a. $T = 20°C$, $T_d = 15°C$ RH = _____%

    b. $T = 20°C$, $T_d = 10°C$ RH = _____%

    c. $T = 0°C$, $T_d = 0°C$ RH = _____%

    d. $T = 35°C$, $T_d = 5°C$ RH = _____%

    e. $T = 0°C$, $T_d = -5°C$ RH = _____%

4) Of the five examples presented in (3), which has the highest overall amount of water vapor in the air? Explain how you know. *Hint: The answer is not c. Relative humidity is related to temperature, so if temperatures are higher and relative humidity is higher, there is more water in the air.*

_____

_____

_____

5) What time of day will generally have the highest relative humidity and what time of day will have the lowest relative humidity? Why?

_____

_____

_____

6) Often members of the public will say, "I observed a temperature of 100°F with 100% relative humidity." Is this statement true? How do you know?

_____

_____

HYDROLOGIC BALANCE AND THE HYDROLOGIC CYCLE

6) The balance between precipitation and evaporation is a delicate one that strongly influences the local climate. Based on what you learned about atmospheric circulations and where the world's deserts are located, describe where on Earth you would expect more evaporation than precipitation on average and where you would expect more precipitation than evaporation.

_____

_____

_____

7) What features in the atmospheric circulation are responsible for the general patterns observed in Figures 7.4 and 7.5? Describe how the two are related to each other.

_____

_____

_____

_____

_____

_____

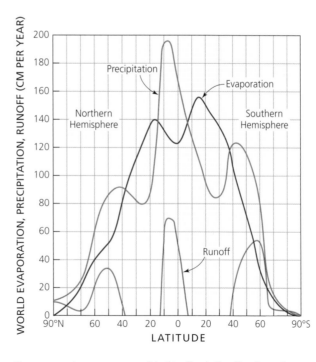

Figure 7.4 Average annual latitudinal distribution of precipitation, evapotranspiration, and runoff in cm per year.

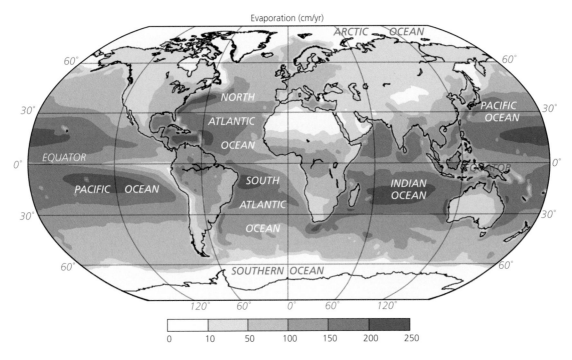

Evaporation (cm/yr)

0	10	50	100	150	200	250	

Figure 7.5 Global distribution of annual evaporation and evapotranspiration in centimeters.

CRITICAL THINKING QUESTIONS

We will now complete the experiment with the requested materials. Pour the hot water into the glass and measure and record its temperature. Add the three ice cubes to the hot water. Once they are added, mark the water level on the outside of the glass using the marker. Wait until all ice has been melted, then measure the temperature of the water again. Mark the water level after the ice has melted as well.

_____ Temperature of hot water _____ Temperature of cooled water

1) Using your understanding of latent heat, list and describe all factors that could have led to the change in temperature you observed in the glass of water. Which process do you think was the most important?

_____

_____

_____

2) What did you notice about the water level after the ice melted? A common misconception about climate change is that melting sea ice will lead to rising ocean levels. After this experiment, do you think that melting sea ice will contribute to rising or sinking ocean levels? Explain your reasoning.

_____

_____

_____

PRECIPITATION, AIR MASSES, AND FRONTS

GOALS OF THIS LAB:

To observe different mechanisms responsible for precipitation formation. To observe the air mass types that affect North America and identify locations that are typically under the influence of each air mass type. To look at different types of fronts.

REQUIRED ITEMS FOR LAB:

Textbook, a globe with topography, an atlas, and Internet access

The midlatitudes mark a zone of distinct contrast between the two dominant *air mass* types in each hemisphere: the warm, humid tropics and the cold, dry polar regions (Fig. 8.1). The position of the jet stream corresponds to the boundary between these two air masses, and they are constantly battling for position (the term "front" is short for "battlefront"). The locations of these fronts are very important, as they mark the locations where one type of air mass clashes with another air mass type and can mark a location of enhanced *frontal lifting.*

The formation of precipitation requires sufficient upward air motion, which results in the formation of large drops or ice crystals that are able to overcome the force of the upward motion and fall toward the surface. This upward motion typically is "forced" by one of four possible most commonly observed mechanisms:

1. *Convergent lifting*: When winds collide near the surface, the air needs to evacuate to make room for more air. Since air cannot go downward into the ground, the only possible solution is for the air to rise. Water vapor in this rising air can easily condense and form clouds and precipitation if the convergence is strong enough. This is a commonly observed phenomenon at the ITCZ, where converging trade winds result in a strong rising motion and heavy precipitation.

2. *Convective lifting*: When the land surface warms to a temperature sufficiently warmer than the air around it, the density will fall to the point that the air rises through convection. Assuming the convective parcel of air continues to remain warmer than the surrounding environment, the air will continue to rise and possibly form clouds and precipitation once the air cools dry adiabatically to its dewpoint and becomes saturated.

3. *Orographic lifting*: If air rises up a physical boundary, such as a mountain, the air will eventually reach saturation as it cools adiabatically and will

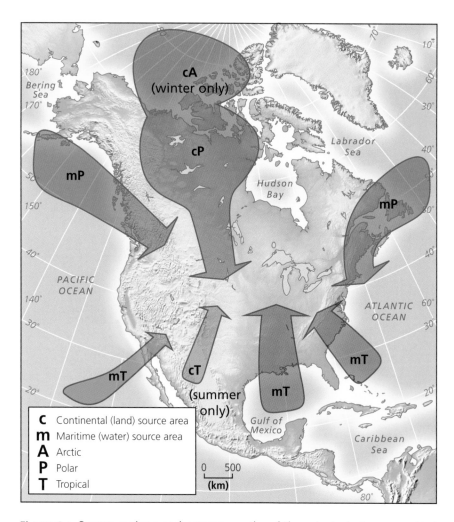

Figure 8.1 Source regions and common paths of the principal air masses that affect the continental United States.

form clouds. If enough rising motion occurs, the clouds will form precipitation. Interestingly, when the air begins to flow down the leeward side of the mountain, the air dries out as it warms adiabatically, forming a rain shadow on the leeward side of a mountain that is often very dry.

4. *Frontal lifting*: The frontal boundary marks a transition zone between two distinct air masses. The warm air at the frontal boundary will rise up over the colder air, meaning any moisture in the warmer air (which often has considerably more moisture than the cold air) will be able to condense and form precipitation.

Knowledge of these lifting mechanisms helps us understand why some locations on Earth receive a significantly large amount of precipitation, while others are relatively dry. Typically, some combination of the factors outlined above is responsible for the precipitation for a region, but occasionally one factor dominates all others.

PRECIPITATION MECHANISMS

For these exercises, you will use the global precipitation patterns provided in Figure 8.2. You should also use a globe with topographic information in answering these questions.

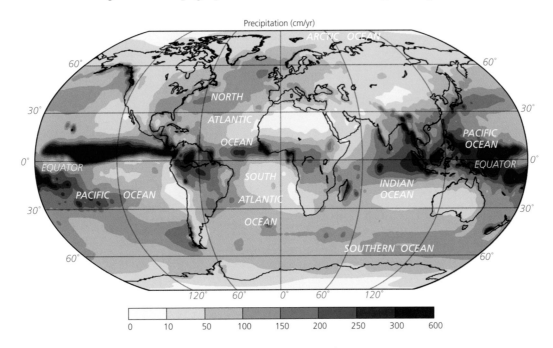

Precipitation (cm/yr)

| 0 | 10 | 50 | 100 | 150 | 200 | 250 | 300 | 600 |

Figure 8.2 Global distribution of annual precipitation in centimeters.

1) What is the primary cause for the low values of precipitation along the west coast of South America? Which precipitation mechanism is most likely at fault?

2) In the mid- to late summer, winds begin to flow onshore along the eastern Indian peninsula. During this season, many locations in India (see the climograph for Trivandrum, India, in Fig. 8.3) receive large quantities of precipitation. In the winter, the winds blow offshore. What lifting mechanisms are in place during the summer (onshore flow) that are absent during the winter (offshore flow)?

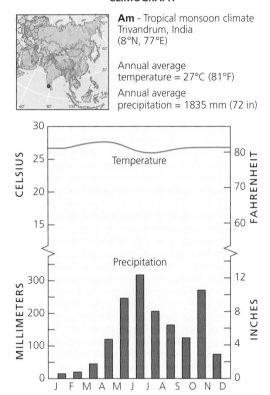

CLIMOGRAPH

Am - Tropical monsoon climate
Trivandrum, India
(8°N, 77°E)

Annual average
temperature = 27°C (81°F)

Annual average
precipitation = 1835 mm (72 in)

Figure 8.3 Climograph of a monsoon rainforest
(Am) weather station.

3) Overall there is a long, somewhat continuous "belt" of heavy rainfall along the equator, yet a relative void of precipitation at the poles. Using the previous discussion on atmospheric circulation, explain why there is such a difference in precipitation in these regions.

_____

_____

_____

4) How might the precipitation map over the southeast United States look if the Gulf of Mexico were not present? Explain.

_____

_____

_____

ASSESSING AIR MASSES AND FRONTS

Use Figure 8.1 to assist in answering these questions.

5) A type of front that separates a cT air mass from an mT air mass is called a *dry line.* The dry line is a very common location for supercell and tornado development, so its position is very important. Based on what you see in Figure 8.1, where is the most common place to observe a dry line in the United States?

_____

_____

_____

6) Typically for the United States, midlatitude cyclones initially develop coming off of the Rocky Mountains and progress eastward toward the East Coast. They typically have a frontal structure that resembles Figure 8.4A. Using Figure 8.4A, provide an example location (e.g., state) that is under the influence of each of the following air masses, and explain how you know this to be true.

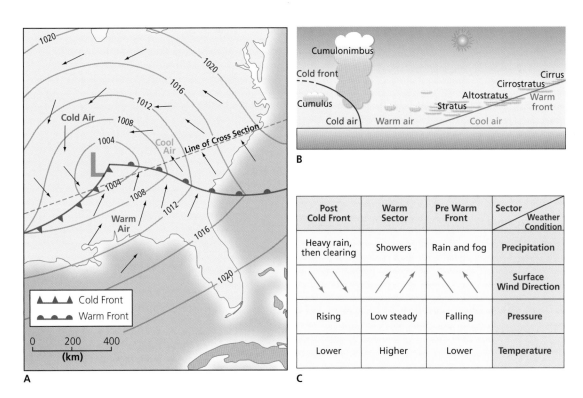

Figure 8.4 Mature stage of a midlatitude cyclone over the southeastern United States. (A) Pressure fields, windflows, and fronts. (B) Cross-sectional view along the dashed line mapped in (A). The vertical scale is greatly exaggerated. Cold fronts typically rise only 1 m for every 70 m horizontal extent; for warm fronts the ratio is about 1:200. (C) Summary of surface weather conditions along the cross-sectional transect.

i) mP

_____

_____

_____

ii) mT

_____

_____

_____

iii) cP

_____

_____

_____

7) State which type of air mass is *most common* in each of these cities:

    i) New Orleans, Louisiana   _____

    ii) New York, New York   _____

    iii) Los Angeles, California   _____

    iv) Mexico City, Mexico   _____

    v) Bismarck, North Dakota   _____

    vi) Seattle, Washington   _____

    vii) Miami, Florida   _____

8) Based on previous discussions on the impact of water vapor on temperature, which air mass type will typically get the coldest? The warmest? Explain how you know.

_____

_____

_____

_____

_____

_____

_____

_____

CRITICAL THINKING QUESTIONS

The map in Figure 8.5 shows specific humidity for the southern Plains on May 3, 1999, the day of an infamous tornado outbreak where the fastest wind speeds ever recorded on the Earth's surface (318 mph) were observed. Specific humidity is a measure of the mass of water vapor (in kg) per mass of dry air (in kg). In other words, for a specific humidity of 0.02 kg/kg, there are 0.02 kg, or 20 g, of water vapor for every 1 kg of air. Thus, higher specific humidity values mean higher moisture. Use this map to answer the following questions:

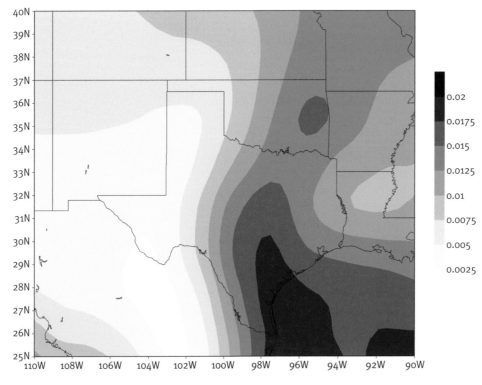

Figure 8.5 Specific humidity (kg/kg) for May 3, 1999, at 1800 UTC.

1) What is the air mass type over eastern Oklahoma and eastern Texas? What is the air mass type over southern New Mexico, northern Mexico, and western Texas?

_____

2) Based on the answer to the previous question, where is the dryline on this image? If the dryline is a focal point for thunderstorm development, where did the tornado outbreak likely occur (list parts of states)?

_____

_____

3) A *dryline bulge* is an area where the dryline "bulges" eastward relative to a traditional north–south orientation that is often observed with a dryline. The dryline bulge is a known area where severe thunderstorm development is highly likely. Do you see a dryline bulge on this image? If you were a forecaster, where would you expect tornadoes to form based on this image? Use the Internet to see where tornadoes actually formed on this day. Were you right?

_____

_____

_____

_____

_____

WEATHER SYSTEMS

GOALS OF THIS LAB:

To understand the basic characteristics of midlatitude cyclones and the air masses and fronts associated with these cyclones. To characterize and summarize weather conditions in different locations surrounding a midlatitude cyclone and understand why those conditions exist.

REQUIRED ITEMS FOR LAB:

Textbook and an atlas

The midlatitudes in each hemisphere, due to their location relative to cold, polar air masses and warm, tropical air masses, represent a "battleground" between these clashing air mass types. Often, storm systems that form in the midlatitudes will form along the boundary between these two air masses, as this boundary also corresponds to the polar front outlined in Lab 6 (see also Fig. 9.1). The *midlatitude cyclone* is characterized by surface winds that are flowing counterclockwise (in the Northern Hemisphere) around a

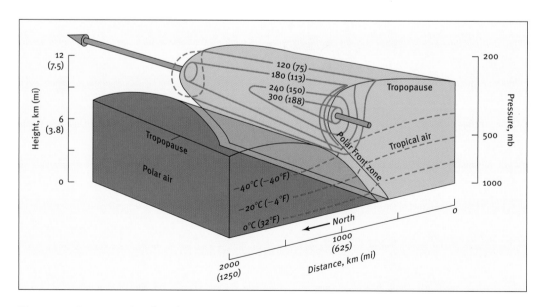

Figure 9.1 Cross-sectional model perspective of the polar front zone and (high-velocity) jet stream over the central United States. The view is from the northwest toward the southeast. The north–south extent of this diagram is roughly the distance from Lake Superior to the Louisiana Gulf Coast. The wind speeds surrounding the jet stream are calibrated in km/h with mi/h equivalents shown in parentheses.

central low pressure. However, they are also directed inward toward the center of low pressure, causing air convergence near the center of the low and rising air (and subsequently precipitation). The unique interaction between the midlatitude cyclone and the polar/tropical air masses that surround it allows for the identification of the following "phases of life" of an extratropical cyclone:

1. Early stage: This stage represents the initiation of the midlatitude cyclone. A center of low pressure has begun to develop due to dynamic processes in the upper atmosphere (e.g., the polar front/polar jet) and winds have begun to circulate around this center of low pressure. Often, this stage is characterized by a *stationary front*.

2. Open wave stage: During this stage, the system has well-defined frontal boundaries and a well-defined warm sector. Warm air is rising up over cold air on the east side of the cyclone, and on the west side, the cold air is pushing the warm air upward (since the cold air is more dense).

3. Occlusion stage: Eventually, the surging cold front will catch up with the slowly advancing warm front, meaning that at the center of the cyclone, the warm air will all be eventually lifted from the surface. This warm air at the surface is the main energy source of the cyclone, so once this occurs, the storm will begin to dissipate. This process of "catching up" is called *occlusion*.

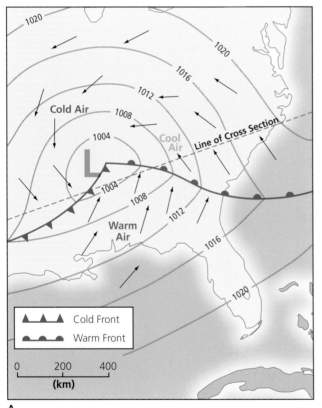

A

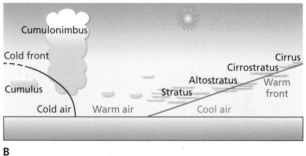

B

Post Cold Front	Warm Sector	Pre Warm Front	Sector / Weather Condition
Heavy rain, then clearing	Showers	Rain and fog	Precipitation
↘ ↘	↗ ↗	↖ ↖	Surface Wind Direction
Rising	Low steady	Falling	Pressure
Lower	Higher	Lower	Temperature

C

Figure 9.2 Mature stage of a midlatitude cyclone over the southeastern United States. (A) Pressure fields, windflows, and fronts. (B) Cross-sectional view along the dashed line mapped in (A). The vertical scale is greatly exaggerated. Cold fronts typically rise only 1 m for every 70 m horizontal extent; for warm fronts the ratio is about 1:200. (C) Summary of surface weather conditions along the cross-sectional transect.

4. Dissipating stage: Once the storm has fully elevated all warm air at the surface, the storm has lost the surface instability and its energy, causing the storm to dissipate. At this point, cold air lies below warm air over the entire cyclone's depth.

Note that as these processes occur, the cyclone *propagates* (moves by reforming) through the motion of the jet stream. While the jet stream can be thought of as a steering current, it is not pushing the cyclone like a stick in a stream; it is dynamically reforming the cyclone as the jet stream itself moves.

The midlatitude cyclone's life phases are associated with very commonly observed weather phenomena. In addition, positions relative to the center of the midlatitude cyclone and the cyclone's fronts have common weather characteristics. Figure 9.2 characterizes these positions within a cyclone and the weather changes associated with each position.

For these exercises, you will be assessing a weather map of the famous March 12–14, 1993, Superstorm that affected the East Coast of the United States. The map in Figure 9.3 will be used in many of your questions. In this map, the solid lines represent isobars in mb, while the dashed lines represent isotherms in K. Use this map to answer the following questions.

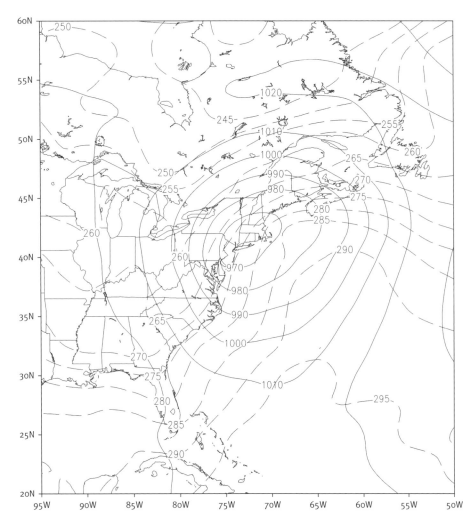

Figure 9.3 Isobars (solid lines) and isotherms (dashed lines) for the Earth's surface on 0600 UTC 14 March 1993.

1) On the map in Figure 9.3, identify the locations of the cold front and the warm front on the midlatitude cyclone. Be sure that the symbols on each frontal boundary point the correct way, and explain your choice of locations.

_____

_____

_____

2) Based on your positioning of the fronts, identify the locations of the continental polar (cP) air mass, the maritime tropical (mT) air mass, and the maritime polar (mP) air mass on your map. Explain briefly the reasoning for your selections.

_____

_____

_____

3) Which direction is this midlatitude cyclone moving? How do you know?

_____

_____

_____

4) What phase of life is this cyclone currently undergoing? Explain how you know.

_____

_____

_____

5) Describe with some detail the weather conditions that are currently occurring in the following locations. You can use an atlas to find these cities.

    i) Nashville, Tennessee

_____

_____

_____

    ii) Orlando, Florida

_____

_____

_____

iii) Boston, Massachusetts

_____

_____

_____

iv) Bangor, Maine

_____

_____

_____

6) Based on what you know about the motion of the storm and the impacts that the mid-latitude cyclone and its fronts will have on future weather, complete a weather forecast for the next 24 hours for the locations in (5) using the space below.

i) Nashville, Tennessee

_____

_____

_____

ii) Orlando, Florida

_____

_____

_____

iii) Boston, Massachusetts

_____

_____

_____

iv) Bangor, Maine

_____

_____

_____

7) Explain the main differences between a midlatitude cyclone such as this event and a tropical cyclone. (If necessary, see pages 157–59 in *Physical Geography*, 4th ed., for an overview of tropical cyclones.) What is present on your map above that would be notably absent from a tropical cyclone?

_____

_____

_____

CRITICAL THINKING QUESTIONS

The process of occlusion refers to the "catching up" of the surging air behind the cold front with the warm front, lifting the warm air in the warm sector upward and effectively cutting off the energy source of the low. However, the air mass characteristics in occluded fronts are very different depending on where the occlusion process takes place (see Fig. 9.4).

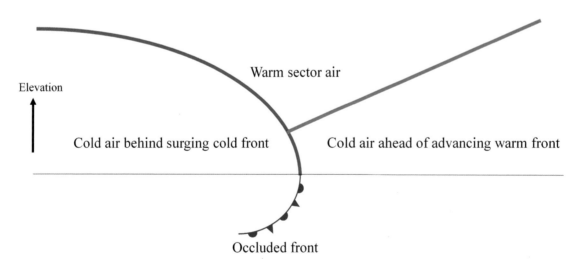

Figure 9.4 Example air mass configuration associated with an occluded front. Note that the two cold air masses originate from different sources and thus have different temperature and humidity characteristics.

It is possible for the air behind the occluded front to be either colder or warmer than the air ahead of it. Such scenarios refer to *cold occlusions* (in the case of colder air replacing the cool maritime air) or *warm occlusions* (in the case of the warmer maritime air replacing cold polar air).

1) Draw pictures similar to Figure 9.4 that represent a *warm occlusion* and a *cold occlusion*.

2) Which type of occlusion is most common in the United States? Why?

_____

_____

_____

3) For whichever type is *not* most common in the United States, where would you expect to find such a situation? Explain your reasoning.

_____

_____

_____

CLIMATE CLASSIFICATION AND REGIONALIZATION

GOALS OF THIS LAB:

To understand the locations of particular climate regimes and the reasons for their existence, and to understand the differences in climates for different regions.

REQUIRED ITEMS FOR LAB:

Textbook, globe, and elevation map/Internet search

The term *climate* refers to "average weather conditions" for a particular location. The Earth's climate system has a very predictable behavior, based on the major climate controls. These controls include:

- latitude
- proximity to water
- ocean circulations
- general circulation
- elevation.

Because the impacts of these controls are fairly well known for different regions on Earth, it is possible to develop a climate classification system that outlines these characteristics and describes local climates. The perfect climate system should accomplish five primary tasks:

1) differentiate the major climates that occur on Earth
2) show relationships among these climates
3) apply to the entire world
4) provide a framework for further subdivision into subclimates within the major climate systems

5) demonstrate the controls that are responsible for the distribution of the climate systems.

The Köppen climate system provides this balanced approach as it provides a single system that describes the entire Earth climate system with relative simplicity. The Köppen system outlines six major climate groups (denoted as letters A, B, C, D, E, and H). The A group (major tropical) corresponds to warm climates with relatively high amounts of precipitation. In the B (dry climate) group, climates are described by a lack of precipitation compared to the atmosphere's demand for water (potential evaporation) and can be warm or cold. The C (mesothermal) climate groups are midlatitude climates that are characterized by a moderate seasonal cycle in air temperature and a wide range of precipitation availability. The D (microthermal) group is representative of a climate group with very strong seasonality in temperature and variable precipitation. With their very cold winter, D climates are therefore a more severe climate than the C group. The E (polar) group is largely represented by low temperatures throughout the year, with either a short summer thaw or year-round ice. The final group, the H (highland) group, represents places where elevation exerts a dominant control over local climate. Climates resembling other categories are often found in the H group, with changes from one group to another occurring over a very small horizontal distance, due to the wide variety of elevations in these regions.

As shown in Figure 10.1, these climate groups are broken down into subgroups (and some into further subgroups) based on the availability of precipitation, differences in temperature, or a combination of these two factors.

Figure 10.2 provides a hypothetical outline of the locations of these different climates, based on their locations on a hypothetical continent.

While the horizontally distributed climate regions are fairly easy to diagnose, all five horizontal zones may exist at the same time in the H climate group. The five climate zones in an area with high elevation are outlined with a slightly different naming convention, based more on the temperature in the region than on the precipitation amount. Figure 10.3 shows how it is possible for all five climate zones to exist simultaneously over a very small (tens of miles) horizontal extent.

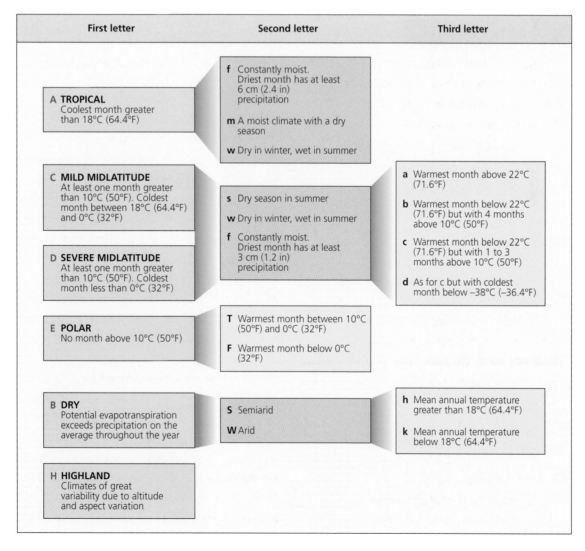

Figure 10.1 Simplified version of the modern Köppen climate classification system.

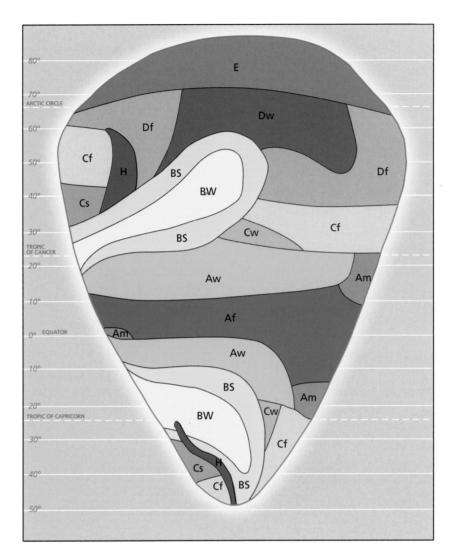

Figure 10.2 Position of different Köppen climate regions on a hypothetical continent that extends through both hemispheres.

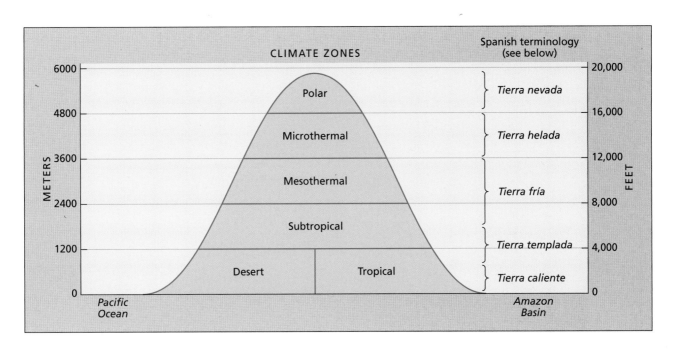

Vertical zone	Elevation range	Average annual temperature range
Tierra nevada	4800+ m (16,000+ ft)	<-7°C (<20°F)
Tierra helada	3600–4800 m (12,000–16,000 ft)	-7°–13°C (20°–55°F)
Tierra fría	1800–3600 m (6000–12,000 ft)	13°–18°C (55°–65°F)
Tierra templada	750–1800 m (2500–6000 ft)	18°–24°C (65°–75°F)
Tierra caliente	0–750 m (0–2500 ft)	24°–27°C (75°–80°F)

Figure 10.3 Highly generalized west-to-east cross-section of the Andes in equatorial South America. The vertical climatic zonation shown on the mountain corresponds to the Spanish terminology listed.

1) Match these 12 locations with their primary climate zone (e.g., mark each location with an A, B, C, D, E, or H).

i) Denver, Colorado _____

ii) San Francisco, California _____

iii) Seattle, Washington _____

iv) Miami, Florida _____

v) Antarctica _____

vi) Mt. Everest _____

vii) New York, New York _____

viii) London, UK _____

ix) Beijing, China _____

x) Sydney, Australia _____

xi) Honolulu, Hawaii _____

xii) Phoenix, Arizona _____

2) Using the three climographs in Figures 10.4, 10.5, and 10.6, determine what the Köppen classification system designation would be. Explain your reasoning for each.

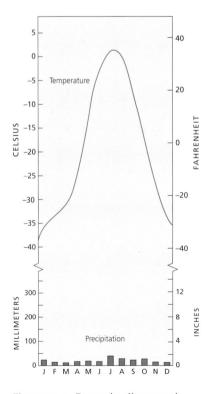

Figure 10.4 Example climograph.

Explanation:

_____

_____

_____

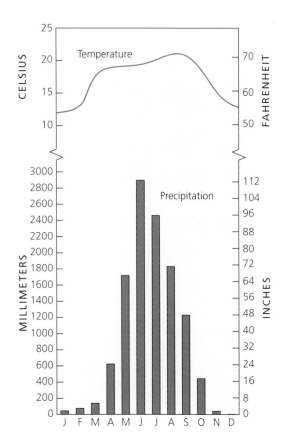

Figure 10.5 Example climograph.

Explanation:

_____

_____

_____

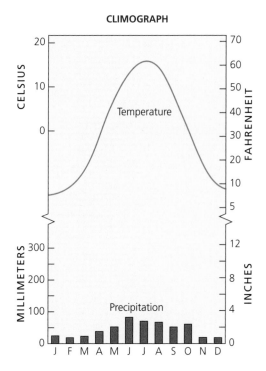

CLIMOGRAPH

Figure 10.6 Example climograph.

Explanation:

_____

_____

_____

3) Use the data and blank climographs in Tables 10.1 and 10.2 and Figures 10.7 and 10.8 to look at the climate at these hypothetical locations. Speculate about the type of climate in place at each location, and if this location exists in the Northern Hemisphere or the Southern Hemisphere. Explain your reasoning.

Temperature (Celsius)	Precipitation (mm)
36	5
31	9
28	11
24	12
21	14
19	15
19	13
22	11
24	9
26	7
27	6
32	4

TABLE 10.1

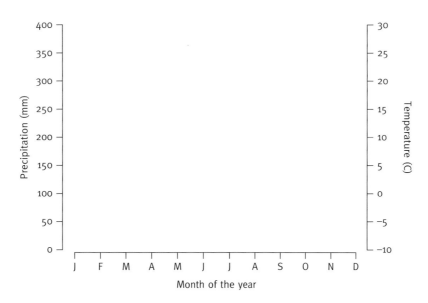

Figure 10.7 Blank climograph.

Explanation:

_____

_____

_____

TABLE 10.2

Temperature (Celsius)	Precipitation (mm)
−1	65
4	80
9	95
16	110
23	135
25	165
24	155
17	105
10	95
6	85
2	65
−3	55

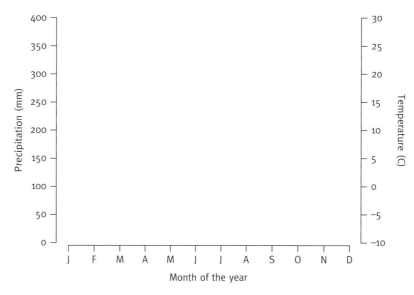

Figure 10.8 Blank climograph.

Explanation:

_____

_____

_____

4) Using an elevation map, estimate the elevation of each of the following mountains. Identify the horizontal primary climate type (A–E) at the peak of each of the following mountains. You should use Figure 10.3 as the basis for your reasoning.

	Elevation	Climate type
i) Mt. Everest	_____	_____
ii) Mt. Washington, New Hampshire	_____	_____
iii) Diamond Head, Hawaii	_____	_____
iv) Mt. Redoubt, Washington	_____	_____
v) Mt. McKinley, Washington	_____	_____

CRITICAL THINKING QUESTIONS

1) The cities of Savannah, GA (32.01°N) and San Diego, CA (32.7°N) have similar latitude, yet vastly different climates. In particular, Savannah is much warmer and gets much more precipitation than San Diego. Considering they have the same latitude and both lie on coasts (San Diego on the West Coast, Savannah on the East Coast), use their local climates to discuss the primary reasons for the major differences in their climates.

_____

_____

_____

Look at Figure 10.9, which summarizes all of the Köppen climate classifications for the entire Earth. Answer the questions that follow.

WORLD CLIMATES
After Köppen-Geiger

A Tropical Climate

Af	No dry season
Am	Short dry season
Aw	Dry winter

B Dry Climate

BS	Semiarid
BW	Arid

h = hot
k = cold

C Mild Midlatitude Climate

Cf	No dry season
Cw	Dry winter
Cs	Dry summer

D Severe Midlatitude Climate

Df	No dry season
Ds	Dry summer
Dw	Dry winter

a = hot
    summer
b = cool
    summer
c = short, cool
    summer
d = very cold
    winter

E Polar Climate

E	Tundra and ice

F Highland Climate

H	Unclassified highlands

0   2000

(km)

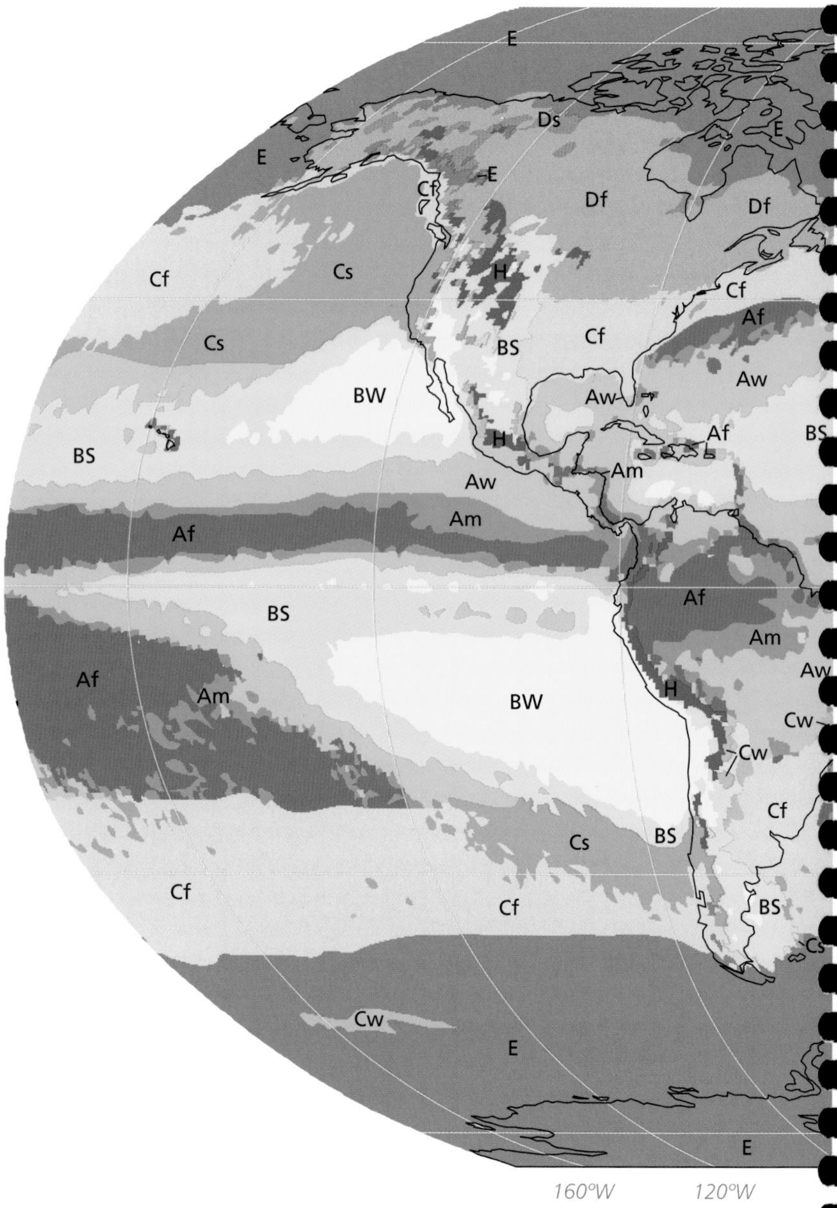

Figure 10.9 Köppen climate classifications for all locations on Earth.

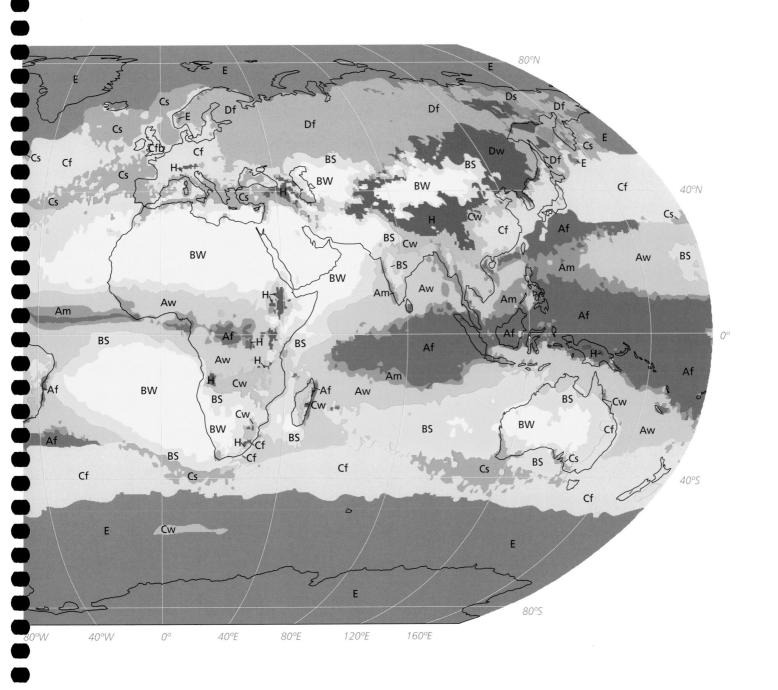

2) Discuss the similarities with eastern North America and eastern Asia. Why do these exist?

_____

_____

_____

3) What are the primary climate types in the nation of Russia? How do these relate to the annual temperature variability observed in Russia?

_____

_____

_____

4) Relate the general patterns you see in the major climate groups in Figure 10.9 with the discussion in the previous lab about the general circulation. How do the major climate regions line up with the major points in the general circulation? Do these relationships make sense? Explain your reasoning.

_____

_____

_____

5) What would the Köppen climate classifications look like if the Earth were a completely water-covered planet?

_____

_____

_____

NATURAL AND ANTHROPOGENIC CLIMATE CHANGE

GOALS OF THIS LAB:

To understand the factors responsible for changes in the Earth's climate. To understand radiation balance and how it contributes to shifts in the Earth's climate. To understand the role of natural and manmade (anthropogenic) processes on climate change.

REQUIRED ITEMS FOR LAB:

Textbook, calculator, and a ruler

While the term *climate* refers to long-term, relatively consistent weather conditions, it is often mistakenly assumed that the Earth's climate remains stationary. There are many factors that contribute to changes in the Earth's climate on various time scales ranging from a few weeks to millennia. The basis for most long-term climatic shifts on Earth goes back to the basic radiation budget that was outlined in Lab 4 (see also Figs. 11.1 and 11.2).

The radiation budget is the sum of radiation gains and losses (as outlined in Lab 4). We can

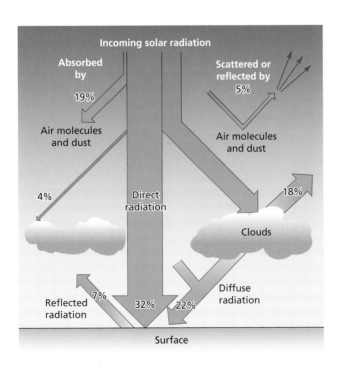

Figure 11.1 Solar radiation flows in the atmosphere.

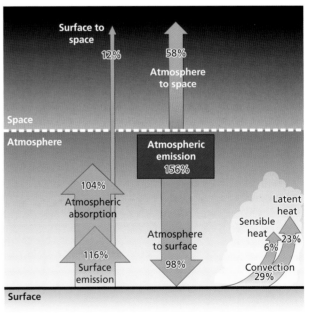

Figure 11.2 Global mean terrestrial radiation and convection. All values are expressed as a percentage of solar radiation entering the top of the atmosphere.

consider the radiation budget of the surface alone or the atmosphere alone, or we can consider a column running from the surface to the top of the atmosphere, in which case we evaluate radiation gains and losses through the top of the atmosphere. The term "budget" is a little misleading, because radiation gains and losses need not be in balance for either the surface or atmosphere at any one location on Earth. Of course, because radiation is the only way energy moves in and out of the Earth's system, gains and losses through the top of the atmosphere must be equal in order to conserve energy. In this lab we will look only at the surface radiation budget, which can be written as

$$R_{net} = S\downarrow - S\uparrow + L\downarrow - L\uparrow$$

The total value R_{net} corresponds to an accumulation of radiation (heat) at a location, so if the value at a location increases, the temperature at that location is likely to increase as well, resulting in local changes in temperature manifest through the surface energy budget (that is, sensible and latent heating). The four terms on the right-hand side correspond to

- $S\downarrow$—incoming solar radiation
- $S\uparrow$—outgoing solar radiation reflected by the Earth's surface (albedo)
- $L\downarrow$—radiation emitted by the atmosphere downward to the surface.

This term is an energy source to the surface and represents the *greenhouse effect*. (From the atmosphere's perspective it is an energy loss.)
- $L\uparrow$—radiation emitted upward by the surface. As you know, most of this is absorbed by the atmosphere and only a small fraction escapes into space.

This equation is helpful for quantifying and understanding climate change, as any changes to these major radiation budget terms represent a change in climate. Indeed, successful modeling of climate change depends in no small part on faithful representation of radiation budgets.

Looking back at the history of the Earth's temperature record over millions of years (using proxy measures such as ice core data) and even the last 150 years of thermometer measurements leaves no doubt that there have been large changes in radiation budgets and climate (see Fig. 11.5 at the end of this lab for approximately the last 150 years). Figure 11.3 represents the global average temperature changes over the last 70 million years, while Figure 11.4 shows the more recently (last 1,000 years) observed increases in temperature. The primary ongoing research activities in climate change center on the causes of these trends in the temperature record.

Many natural and anthropogenic (human-caused) factors can contribute to changes in

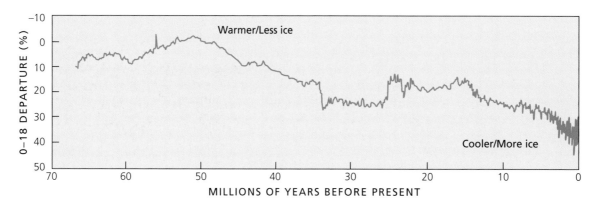

Figure 11.3 Climate of the last 70 million years as indicated by oxygen isotope records. The curve reflects changes in both ice amount and temperature.

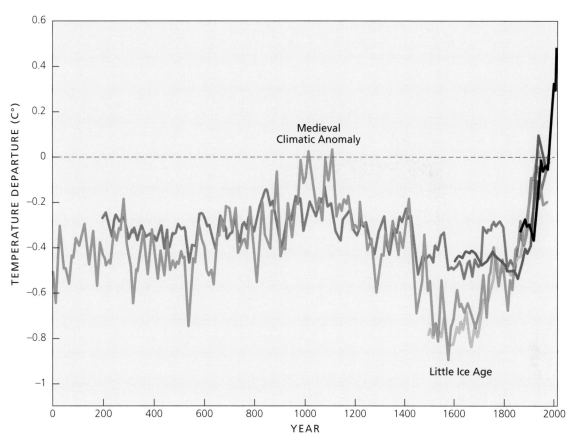

Figure 11.4 Temperature departures from the long-term mean for the last 2,000 years as estimated by various proxy measures (colored curves) and by thermometers (black) since 1854.

the radiation budget provided above, resulting in shifts in radiation at a particular location and changes in local and global temperatures. Some of these factors include:

- volcanic eruptions
- introduction of pollutants into the atmosphere
- release of greenhouse gases through burning of fossil fuels
- shifts in the Earth–Sun relationships
- changes in cloud cover
- changes in the Sun's radiation output

- human-caused modification of the land surface
- natural changes in the land cover of Earth (modifying the albedo)
- melting or freezing of polar ice, which modifies ocean depths and global albedo
- urbanization.

A primary challenge of the climatologist is to isolate the impacts of these changes and determine how they may result in a change in radiation budgets and global temperature.

Based on the above discussions on climate change, answer the following questions.

1) Of the 10 climate change factors listed, identify which are anthropogenic and which are natural (circle the correct response).

- Volcanic eruptions

  natural/anthropogenic

- Introduction of pollutants into the atmosphere

  natural/anthropogenic

- Release of greenhouse gases through burning of fossil fuels

  natural/anthropogenic

- Shifts in the Earth–Sun relationships

  natural/anthropogenic

- Changes in cloud cover

  natural/anthropogenic

- Changes in the Sun's radiation output

  natural/anthropogenic

- Human-caused modification of the land surface

  natural/anthropogenic

- Natural changes in the land cover of Earth (modifying the albedo)

  natural/anthropogenic

- Melting or freezing of polar ice, which modifies ocean depths and global albedo

  natural/anthropogenic

- Urbanization

  natural/anthropogenic

2) Using the climate change factors listed above, determine if each of these factors would lead to a net increase, decrease, or no change in the budget equation. Explain your reasoning.

_____

_____

_____

3) You'll notice that the terms with the $\downarrow$ arrow are positive in the budget equation. That means they are incoming toward the surface. Accordingly, the value of R_{net} will have a positive or negative sign, and that sign will correspond to whether more radiation is being absorbed at the surface (positive sign) or escaping from the surface (negative sign). Based on this reasoning alone (i.e., ignoring sensible and latent heating), determine if the net radiation will lead to a net warming at the surface or a net cooling for each of these scenarios (circle one):

    i) $S\downarrow = 400$ W/m²; $S\uparrow = 150$ W/m²; $L\downarrow = 150$ W/m²; $L\uparrow = 250$ W/m²
    warming/cooling

    ii) $S\downarrow = 350$ W/m²; $S\uparrow = 250$ W/m²; $L\downarrow = 175$ W/m²; $L\uparrow = 450$ W/m²
    warming/cooling

    iii) $S\downarrow = 0$ W/m²; $S\uparrow = 0$ W/m²; $L\downarrow = 250$ W/m²; $L\uparrow = 350$ W/m²
    warming/cooling

    iv) $S\downarrow = 0$ W/m²; $S\uparrow = 0$ W/m²; $L\downarrow = 150$ W/m²; $L\uparrow = 250$ W/m²
    warming/cooling

4) Using the scenarios provided in (3), describe the radiation characteristics of each scenario by answering the questions below. Explain your reasoning.

    i) Which scenario(s) are likely cloudy?

_____

_____

_____

    ii) Which scenario(s) are likely at night?

_____

_____

_____

iii) Which scenario(s) have a warmer Earth surface?

_____

_____

_____

iv) Which scenario(s) have a larger solar output?

_____

_____

_____

5) The theory of global warming deals directly with the introduction of anthropogenic greenhouse gases into the atmosphere, thereby increasing the impact of the greenhouse effect. Which term would likely increase in the radiation budget equation assuming an increasing greenhouse effect? Based on the equation, would this lead to a net warming or cooling of the global temperature? Explain.

_____

_____

_____

6) This lab has focused primarily on radiation as a driver of climate change on long time scales. Describe some examples of shorter-term climate shifts that may not be directly related to changes in the radiation budget.

_____

_____

_____

CRITICAL THINKING QUESTIONS

The ongoing debate regarding the recent (last 150 years or so) increase in global temperature as a human-caused problem, the theory of *global warming*, has met with resistance due primarily to political and nonscientific issues. However, the current methods by which global temperature has been recorded have been called into question. Figure 11.5 shows two time series of global annual

HadCRUT4 Global Average Temperature Measurements

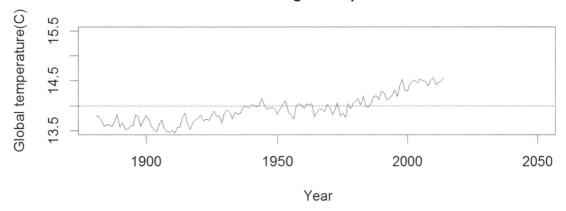

NASA GISTEMP Global Average Temperature Measurements

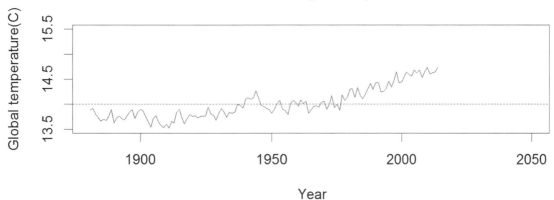

NASA GISTEMP minus HadCRUT4 Global Average Temperature

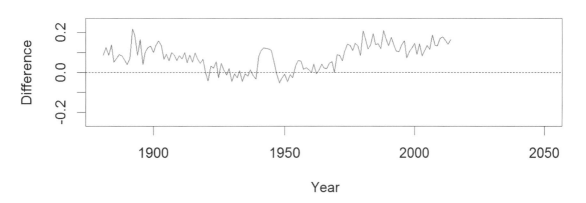

Figure 11.5 HadCRUT4 annual global temperature data (top panel) and NASA GISS global temperature data (middle panel), as well as their mathematical difference.

average temperature, as well as a third time series representing the difference between those two time series. Some details about the datasets:

- The HadCRUT4 dataset originated from the Met Office Hadley Centre of the United Kingdom. The data consist of an ensemble of 100 global land and sea surface gridded realizations and include information about uncertainty in the data. The gridded data came about from a reanalysis of station observation data from across the world and thus are loosely based on a climate model. The data provided in the figure represent the median dataset from the 100 realizations.
- The NASA GISS dataset is formulated using station observations corrected to a 8,000-grid box grid based on a reference grid value for each grid box. These observations are averaged in the east–west direction, corrected for sea surface temperatures, and these east–west averages are averaged to obtain a global average.

Use these datasets to answer the following questions.

1) The average denoted by the dashed line in the first two time series is 14°C. During which decades was the temperature below average? Above average? Was it consistent with both datasets?

_____

_____

_____

2) Use your ruler to draw a line that roughly fits the trend of the first two time series. Extend your line out to 2050. What is your projected temperature for 2050, assuming the trend continues as is?

_____

_____

_____

3) Now, use the ruler to extend the trend starting in 2000 instead of the beginning of the time series. Do you get the same temperature? Is it higher or lower? What does this say about the uncertainty in the change in temperature between now and 2050?

_____

_____

_____

4) The third graph represents the temperature as measured by NASA GISTEMP minus the temperature measured by the Hadley Centre. What do you notice about the differences between these two datasets? How do those numbers compare with the total changes in temperature that have been observed over the full 150-year period? Comment on the implications of this.

SOILS

GOALS OF THIS LAB:
To gain an understanding of the formation, composition, structure, and classification of soils.

REQUIRED ITEMS FOR THIS LAB:
Textbook

Soils, like the air we breathe and the water we drink, are an often overlooked and underappreciated resource. Our very survival is literally rooted in the soil that we walk, drive, and build upon. Soil is the basis for the majority of the world's agriculture, and therefore it produces most of the world's food supply. The study of soils is a relatively young academic discipline, having emerged as an independent area of study from its original discipline of geology. This is an entirely logical transition, because the majority of materials that make up what we know as soils began as rocks. From the moment of their formation, rocks are immediately attacked by the forces of nature that constitute the process we know as *weathering*. Weathering processes fall into two general categories: *physical weathering*, also sometimes referred to as mechanical weathering, and *chemical weathering*. The breakdown of rocks into smaller particles produces the components of what we know as soils. As the text points out, the processes that form soils have many variables and are best expressed through a simple equation, $S = Cl, O, R, P, T$. Soils are the product of a combination of these factors:

climate (Cl), organics (O), relief, also referred to as topography (R), parent material (P), and time (T). In this instance, time refers to geologic time, as the process of soil development is usually a long and gradual one. Because of the number of variables that impact soil formation, it is understandable that there are many different kinds of soils found around the world.

To gain a better understanding of soils, it is important to learn the basic structural and compositional aspects of the majority of the world's soils. To accomplish this, we will start with a few fundamentals on the general structure of soils.

Soil development, as previously stated, is a long-term process, and soils form and evolve over time. As a result of this evolution, soils can be organized according to their degree of development, ranging from weakly developed soils to well-developed soils. Part of the criteria used to classify soils by development is the degree to which a process called *horizonation* has taken place. Horizonation means that through the processes described in the text, the soil particles, minerals, and organics have settled into layers, or *horizons*. A soil that is described as a well-developed soil has achieved what is called a fully horizonated state. This forms what is known as a *soil profile*. Use Figure 12.1 and the text to answer the following questions.

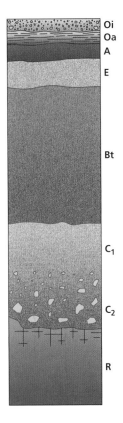

Figure 12.1 Soil profile typical of the humid midlatitudes showing the various soil horizons.

1) Which horizons make up the zone of eluviation, also referred to as the zone of leaching?

_____

_____

_____

2) Which horizons make up the zone of illuviation, also referred to as the zone of accumulation?

_____

_____

_____

3) What is the purpose of designations such as Bt, Oa, or Bw on a soil profile?

_____

_____

_____

4) What is the "O" horizon, and does every soil have one?

_____

_____

_____

5) Which horizons constitute what is known as topsoil?

_____

_____

_____

Soils comprise broken fragments of rock called *detritus*. These fragments come from a wide variety of rocks and are produced from a combination of both physical and chemical weathering processes. They can be organized into the three primary components that make up a soil: sand, silt, and clay particles. One method of classifying soils is by analyzing the percentage of each of these materials in a given sample of soil. This type of soil classification, known as the *textural* classification of a soil, is a relatively simple way to differentiate between soils. Textural classification refers not only to the type of material that makes up a soil but the size of that material as well, with sand grains being the largest, followed by silt grains, and with clay particles the smallest. Farmers are especially interested in knowing the textural classification of a soil because it plays a major role in the fertility, drainage capacity, and moisture retention properties of a soil, all of which have an impact on crop yields. Soil texture can be determined by using the United States Department of Agriculture (USDA) Soil Textural Classification Triangle.

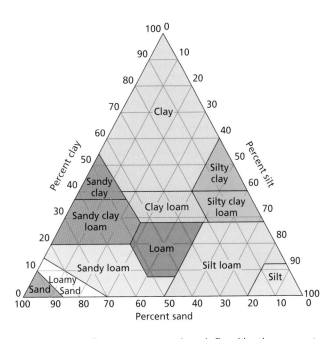

Figure 12.2 Soil texture categories, defined by the percentages of sand, silt, and clay found in a soil sample.

As you can see in Figure 12.2, the classification triangle is shaped like a pyramid, with each of the sides corresponding to sand, silt, and clay. To utilize the chart, a farmer would take a soil sample from a field and have an analysis done. This analysis would give the percentage of sand, silt, and clay in the sample, which would be representative of that particular field. To find the textural soil type of that field, the farmer would take the percentage of each component and locate that percentage on the correct face of the pyramid. Where the three lines intersect will give you the general textural soil type, such as a loam, a silt loam, a sandy loam, and so forth.

Using the chart, find the correct textural soil type for the following compositions:

6) 40 percent sand, 40 percent silt, 20 percent clay

7) 60 percent sand, 10 percent silt, 30 percent clay

_____

8) 30 percent sand, 40 percent silt, 30 percent clay

_____

9) 10 percent sand, 70 percent silt, 20 percent clay

_____

10) 80 percent sand, 10 percent silt, 10 percent clay

_____

The other method of classifying soils is through the multitude of characteristics that combine to produce a unique and distinct soil. This is known as the *taxonomic* classification of soils. In order to fully classify a soil taxonomically, a soil must be thoroughly analyzed so that every aspect of its formation and composition is known. As the text illustrates, taxonomic classification is broken down into six categories in order to classify a soil. This process is a step system from the broadest category, the soil order, down to its most precise category, the soil series. When a soil has been classified as a series, it is unique, and while it may have many soils related to it taxonomically, it is the only soil that possesses those specific characteristics. All of the world's soils belong to one of the most generalized category, that of the soil orders. There are currently twelve soil orders in the system:

Entisols	Spodosols
Inceptisols	Histosols
Vertisols	Andisols
Mollisols	Gelisols
Alfisols	Ultisols
Aridisols	Oxisols

Use the world soil map and materials from *Physical Geography*, 5th ed., to answer the following questions.

11) Which soil orders are found in large areas of the world's desert regions?

_____

_____

_____

12) Which soil order is a weakly developed soil that has not yet formed a distinct B horizon?

_____

_____

_____

13) Which soil order, found almost exclusively in tropical rainforest regions, is highly weathered and has a low fertility when cultivated?

_____

_____

_____

14) What is the primary soil order found in the southeastern coastal plain of the United States and is made up of soils once used to grow cotton?

_____

_____

_____

15) What natural physical phenomenon is associated with the Gelisol order?

_____

_____

_____

16) In order to fall into the Vertisol order, a soil must comprise a minimum of how much clay?

_____

_____

_____

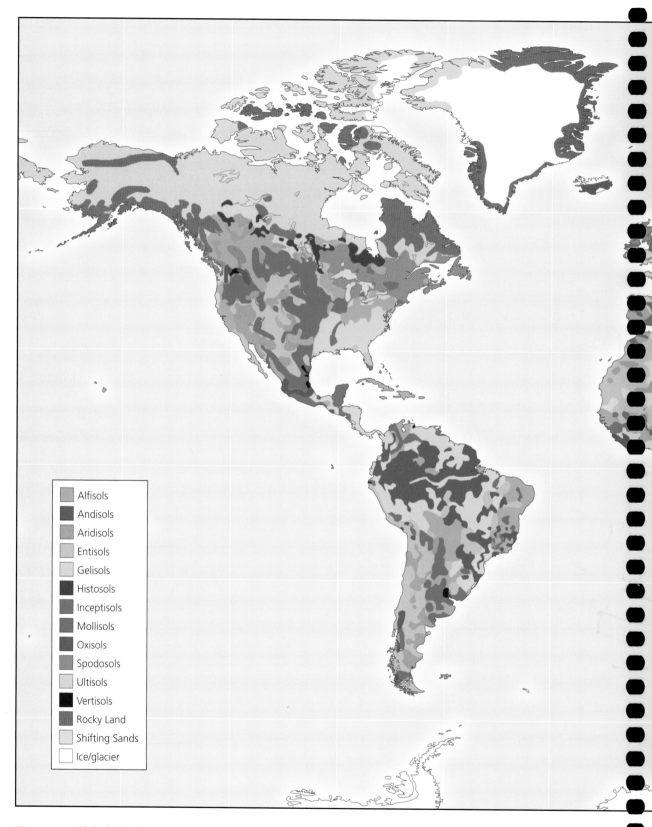

Figure 12.3 Global distribution of soils.

Alfisols
Andisols
Aridisols
Entisols
Gelisols
Histosols
Inceptisols
Mollisols
Oxisols
Spodosols
Ultisols
Vertisols
Rocky Land
Shifting Sands
Ice/glacier

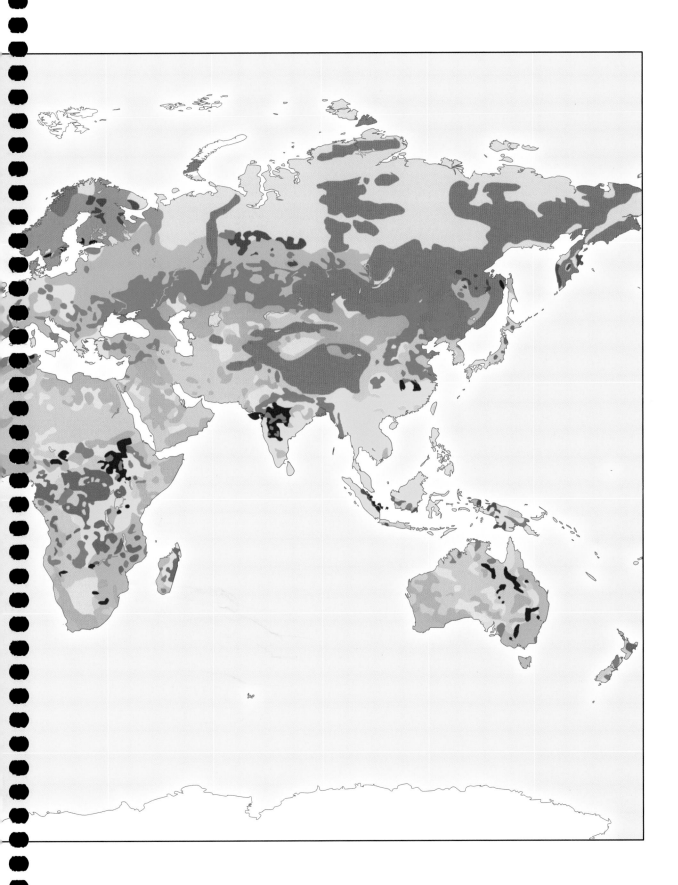

17) Soils that contain a large amount of partially decomposed organic material, as much as 50 percent or more, would fall into which order?

_____

_____

_____

18) The major wheat-producing region of the U.S. Great Plains is situated on soils primarily from which order?

_____

_____

_____

19) Soils that have ashen-gray E horizons, are acidic, and tend to be found in sandy parent materials under high-latitude coniferous forests fall into which soil order?

_____

_____

_____

20) Which soil order contains soils of volcanic origins?

_____

_____

_____

21) Using the world soil map, which soil orders cover the most surface area of the world?

_____

_____

_____

CRITICAL THINKING QUESTIONS

1) What have been some of the impacts of agricultural practices on soils? Be sure to fully explain any negative and positive aspects that you can think of, and give examples for each.

_____

_____

_____

2) How do soils and their characteristics have an impact on the economies and quality of life in nations throughout the world?

_____

_____

_____

ROCKS AND MINERALS

GOALS OF THIS LAB:

To gain an understanding of how rocks and minerals form the Earth's crust.

REQUIRED ITEMS FOR THIS LAB:

Textbook, a laboratory kit for minerals, and each of the three rock classes, if available

Physical geography focuses on the physical features and processes on the surface of the planet. Geology is the study of the structure of the Earth. These two academic disciplines merge in the study of the Earth's crust. The crust and the upper mantle form a region known as the *lithosphere*, which is one of the major spheres that make up the Earth. The crust is the rigid outer portion upon which we live and comprises a combination of the three rock classes, igneous, sedimentary, and metamorphic. These rock classes are in turn made up of *minerals*. While minerals are the building blocks of rocks, the elements are the building blocks of minerals. All matter that we know of is composed of 92 naturally occurring elements. Thus, all minerals are made up of these same elements. Elements form minerals through the chemical process known as *bonding*. There are a few minerals that are composed of a single element, but these are much rarer than those that are made up of two or more bonded elements, which join together to form a *compound*. Consequently, most minerals are compounds of various elements. While there are 92 naturally occurring elements

found in nature, not all are found in equal amounts. In fact, most of the Earth's crust, and therefore the minerals that make up the rocks that represent the majority of the crust, are made up of eight common elements. These eight are the most abundant elements by volume in the Earth's crust. They are:

oxygen (O)   aluminum (Al)
silicon (Si)   potassium (K)
iron (Fe)   calcium (Ca)
magnesium (Mg)   sodium (Na)

MINERALS

By geologic definition, a mineral is a solid, inorganic substance with an orderly internal crystalline structure, definite chemical composition, and unique physical properties. Because of the number of elements and the many possible combinations of these elements, there are thousands of known minerals. Because there are so many minerals, it is inevitable that many of them are similar in color and weight, which means it is sometimes very difficult to differentiate between them. In order to help determine the differences between minerals, geologists developed a system for examining the physical properties of minerals. The text goes into these properties in detail. Use the textbook and answer the following questions on minerals.

TABLE 13.1 Mohs Hardness Scale

Mineral	Hardness
Diamond	10
Corundum	9
Topaz	8
Quartz	7
Potassium feldspar	6
Apatite	5
Fluorite	4
Calcite	3
Gypsum	2
Talc	1

1) What is streak?

_____

_____

_____

2) What is cleavage?

_____

_____

_____

3) What is fracture?

_____

_____

_____

4) How useful is color in differentiating between minerals?

_____

_____

_____

Hardness is an important tool for identifying minerals. The Mohs hardness scale (see Table 13.1) is used to help categorize minerals by their hardness. Use the scale to answer the following questions.

5) What is the hardest mineral?

_____

_____

_____

6) Where does the common silicate mineral quartz rank on the scale?

_____

_____

_____

7) Which is harder, topaz or fluorite?

_____

_____

_____

8) Between which two minerals on the scale does a copper penny fall?

_____

_____

_____

9) Between which two would a pocket knife blade fall?

_____

_____

_____

10) Is color a very useful tool in diagnosing different minerals?

_____

_____

_____

11) Which mineral is more susceptible to the forces of chemical weathering, quartz or calcite, and why?

_____

_____

_____

IGNEOUS ROCKS

The first of the rock classes to form on Earth were those rocks that literally came from fire, _igneous_ rocks. As the molten material, _magma_, from the Earth's mantle begins to cool, both within the crust and on the Earth's surface, chemical bonding begins to take place and leads to the crystallization process, which in turn results in the formation of minerals. The minerals combine as they form to produce the various types of igneous rocks. Use the textbook and answer the following questions.

12) What is the difference between an extrusive igneous rock and an intrusive igneous rock?

_____

_____

_____

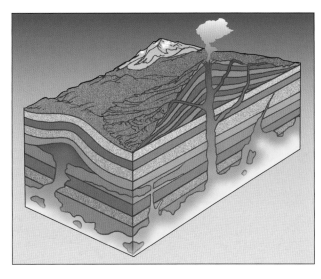

Figure 13.1 Diagrammatic cross-section through the uppermost crust, which shows the various forms assumed by igneous intrusions.

13) What are intrusive igneous bodies called?

_____

_____

_____

Use Figure 13.1 to answer questions 13 and 14.

14) Label the batholith in the illustration.

_____

_____

_____

15) Label the laccolith in the illustration.

_____

_____

_____

16) Give an example of a common igneous mafic rock.

_____

_____

_____

17) Give an example of a common igneous felsic rock.

_____

_____

_____

Figure 13.2 Rock strata on cliff face near Boscastle, Cornwall, England.

SEDIMENTARY ROCKS

The next rock class is *sedimentary.* As weathering wears down existing rock, it is broken down into fragments, which are then moved along the Earth's surface by forces such as wind, water, and glacial ice. Over time, these deposits of broken materials become buried deeper and deeper until the process of compaction and cementation takes place. When this occurs, these former individual grains become pressed together to form sedimentary rocks. There are two general types of sedimentary rock. Those that form from these small particles of broken material are called *clastic.* Sandstone is an example of a clastic rock. Those that form through chemical processes such as solution and organic deposition are called *nonclastic.* Limestone is one example of a nonclastic rock. Sedimentary rocks can be important economically because they are associated with fossil fuel deposits, such as coal, oil, and natural gas. Natural gas and oil can be found in pockets between sedimentary layers called traps and within rocks such as shale. Coal is usually found between strata in layers called seams but sometimes can be near or at the surface.

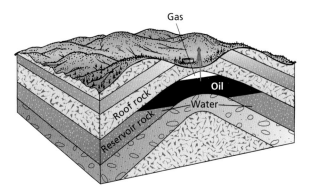

Figure 13.3 Oil and gas traps in sedimentary rock formations.

Figure 13.4 Chalk is a biochemical form of limestone and is a nonclastic sedimentary rock. This formation is part of the cliffs at Beachy Head, Kent, England.

18) Name three common clastic sedimentary rocks.

_____

_____

_____

19) What is a common nonclastic sedimentary rock?

_____

_____

_____

20) Briefly explain the relationship between sedimentary rocks and fossil fuels.

_____

_____

_____

21) What are fossils and how do they form?

_____

_____

_____

METAMORPHIC ROCKS

The last of the rock classes is _metamorphic_ rocks. This class derives its name from the word "metamorphism," meaning to change form. When metamorphism occurs, the existing rock is changed physically, chemically, and structurally. The two primary agents of metamorphism are heat and pressure. Some good examples of metamorphic rocks that resulted from the change of sedimentary rocks would be slate and marble. Slate is shale after it undergoes metamorphism. Marble is limestone after it is put under heat and pressure. Use the textbook and answer the following questions.

22) What is the difference between a foliated and nonfoliated metamorphic rock?

_____

_____

_____

Figure 13.5 Chevron folded metamorphic rock formation. Cornwall, England.

23) Give two common examples each of foliated and nonfoliated metamorphic rocks.

_____

_____

_____

24) What do you call a metamorphic rock formation that covers an extensive area of the Earth's surface?

_____

_____

_____

25) Using rock and mineral specimens, complete the following exercise.

    a. Examine, identify, and describe the characteristics of five common minerals.

_____

_____

_____

b. Examine, identify, and describe the characteristics of an extrusive igneous rock and an intrusive igneous rock.

_____

_____

_____

c. Examine, identify, and describe the characteristics of a clastic and nonclastic sedimentary rock.

_____

_____

_____

Figure 13.6 Stob Dearg, one of the mountains at Glen Coe, Scotland. It is composed primarily of the igneous rock rhyolite and has been heavily weathered by running water and multiple periods of glaciation.

d. Examine, identify, and describe the characteristics of a foliated and a nonfoliated metamorphic rock.

_____

_____

_____

CRITICAL THINKING QUESTIONS

1) How did rocks and minerals help geologists establish the geologic time chart?

_____

_____

_____

2) Explain the difference between mafic and felsic igneous rocks and give two examples of each.

_____

_____

_____

PLATE TECTONICS

GOALS OF THIS LAB:

> To understand how the motion of the Earth's crust has changed the surface of the planet and shaped the world we live in.

MATERIALS REQUIRED FOR THIS LAB:

> Textbook and a world map

When we look at a map of the world, we often take for granted that the Earth's continents and ocean basins have always been as they appear now. This was the assumption for most of human history until relatively recently. In

the early part of the twentieth century, a German geographer named Alfred Wegener was studying climate changes that he believed had taken place throughout Earth's history. In the course of his investigations, he began to

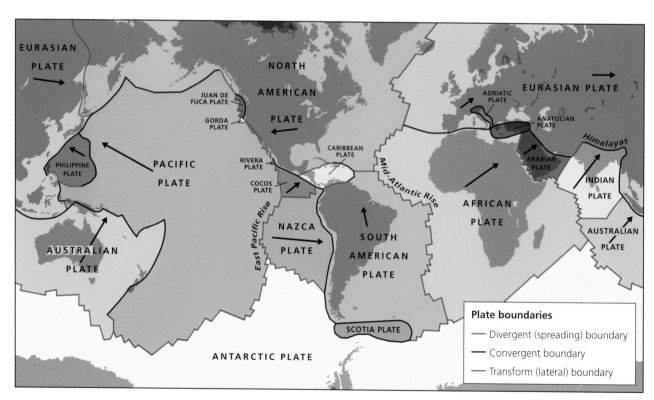

Figure 14.1 Lithospheric plates of the Earth. The relative movement of the two plates at each boundary falls into one of three categories: divergent (spreading), convergent, or transform (lateral) motion. The arrows on the plates show motion relative to a single fixed reference point (maps of plate motion will differ greatly depending on the reference point that is chosen).

notice similarities in rock strata and in the fossil record from different parts of the world. After further investigation, which included physically removing the continents from a map and observing that they joined together into one landmass very neatly, he determined that long ago there had not been separate continents but instead one massive continent, which subsequently broke apart and spread around the Earth. He named this super-continent *Pangaea*, and thus was born Wegener's theory of continental drift. While his theory gained some support within the scientific community at the time, many were skeptical of a theory that stated that the crust was floating around on the Earth's surface constantly changing positions. These skeptics wanted to know exactly how this took place, or what could cause such a process to happen. Unfortunately, Dr. Wegener would never find the force that powered his theory, because the technology of his time was not yet well enough developed, and because he would perish while on an expedition to Greenland in 1930.

Many years later, in the 1960s, the mechanism would be found in the form of massive currents of slowly flowing rock, called *convection cells*, that circulate within the Earth's mantle. Because the Earth's crust is fractured, it is divided into sections called *tectonic plates.* Where these plates meet is known as *plate boundaries.* When flowing rock moves upward from the mantle, magma erupts along these boundaries and forms new crust, and the tectonic plates are pushed away on either side. As they move apart, they slide alongside other plates or directly collide or *converge* with them. The crustal rock near any plate boundary is under stress and can fracture and move along faults, which results in earthquakes. Volcanic activity is common where plates converge or diverge. Perhaps the best example of this is the circum-Pacific belt, which surrounds the Pacific plate. The Pacific plate has

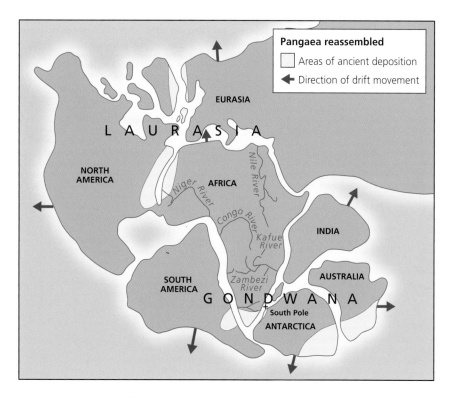

Figure 14.2 The breakup of the supercontinent Pangaea began more than 100 million years ago. Note the radial movement of its remnants away from Africa and how areas of ancient deposition help us understand where today's landmasses were once joined together.

so much volcanic activity that it is known as the "Ring of Fire." While this region is noted for its high activity, this same activity takes place in many other parts of the world as well. Since the Earth is a sphere with a finite, segmented surface area, motion that occurs at one plate boundary can cause a chain reaction that will result in motion in other plates.

There are three basic types of plate boundaries found on Earth's surface:

1. Convergent boundaries where plates move toward one another. Because there are two primary types of crust on Earth, oceanic or basaltic, and continental or granitic, there are three possible types of convergent boundaries. There can be convergence between two continental plates, between two oceanic plates, and there can be a collision between a continental plate and an oceanic plate.

2. Divergent boundaries where plates move apart. These boundaries originate where a plate breaks apart to form a rift. The Mid-Atlantic Ridge is an example of a divergent boundary that was once a rift where what we know as the Western Hemisphere separated from the Eastern Hemisphere and continues to expand.

3. Transform boundaries are plates that grind past one another. An example of this is the boundary between the North American plate and the Pacific plate. The San Andreas Fault is the result of forces associated with this type of boundary.

Convergence between continental plates will result in a collision. This term accurately describes the intense deformation of the plates that occurs there, causing the uplifting of mountain ranges and other surface features. Convergence between oceanic plates can result in the formations of low mountain chains and trenches on the ocean floor. The convergence of continental and oceanic plates will result in the thicker, less dense granitic continental crust overriding the thinner, denser basaltic

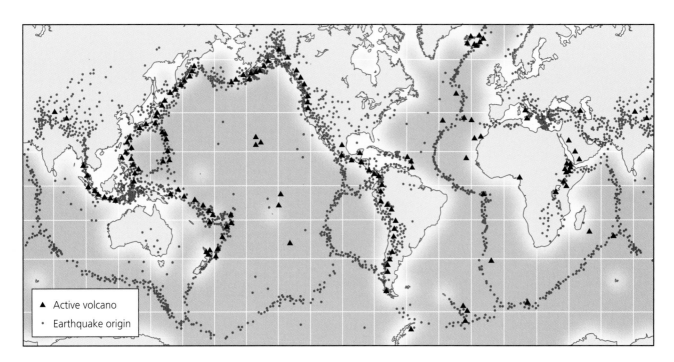

Figure 14.3 Global distribution of recent earthquakes and active volcanoes.

oceanic crust, forcing the oceanic crust into the mantle, known as *subduction*. Therefore, anywhere this type of convergence occurs is called a *subduction zone*. Two prominent features associated with a subduction zone are a mountain range along the coast of the continental plate and an oceanic trench along the coastline. Subduction zones also develop where two oceanic plates converge, as one plate that is denser sinks below the other one.

Subduction is an important aspect of the geologic rock cycle because as new crust is formed somewhere on the Earth's surface as a result of a divergent boundary, such as the Mid-Atlantic Ridge, some of the existing crust must in turn be forced back into the mantle. It is this constant cycle through geologic time that continuously changes the arrangement of the Earth's continents and creates and destroys surface features.

EXERCISES

1) Using the tectonic map in this exercise, which is the world's largest oceanic plate?

_____

_____

_____

2) Using the same map, which is the largest continental plate?

_____

_____

_____

3) Using a physical map of the world, find two examples of a subduction zone and explain why they are occurring in each location.

_____

_____

_____

4) What is the Mid-Atlantic Ridge?

_____

_____

_____

5) Explain what is happening with the Pacific plate.

_____

_____

_____

6) What does the future hold for the Nazca plate and why?

_____

_____

_____

7) Why did Japan recently suffer a massive earthquake?

_____

_____

_____

8) Using the tectonic map above, list the tectonic plates of the world.

_____

_____

_____

9) Using your map, locate and list an example of each of the three types of plate boundaries.

_____

_____

_____

CRITICAL THINKING QUESTIONS

1) If the Mid-Atlantic Ridge continues to build seafloor at its present rate, what would be some of the eventual consequences on the current tectonic plate map? Give specifics in your answer.

_____

_____

_____

2) Explain the processes that created the East African Rift Valley complex and explain what will occur if those processes continue and why.

_____

_____

_____

VOLCANOES

GOALS OF THIS LAB:

To understand volcanoes and volcanic processes.

REQUIRED FOR THIS LAB:

Textbook and a geopolitical map of the world

The tectonic plates that were discussed in the previous unit rest on top of the Earth's mantle. Within the mantle is molten pressurized *magma* that rises toward the surface, often near divergent or convergent plate boundaries. Magma can rise through the mantle in several different ways, depending on the type of boundary, and reach the surface and erupt to form a volcano. Volcanoes occur most commonly where plates diverge and converge but can be found far away from plate boundaries as well. A good example is the highly volcanic region that surrounds the Pacific plate, known as the circum-Pacific belt, also known as the "Ring of Fire." Much of this area is a massive subduction complex and is highly volcanic. There are many volcanoes in the Andes Mountains along the western margin of the South American plate,

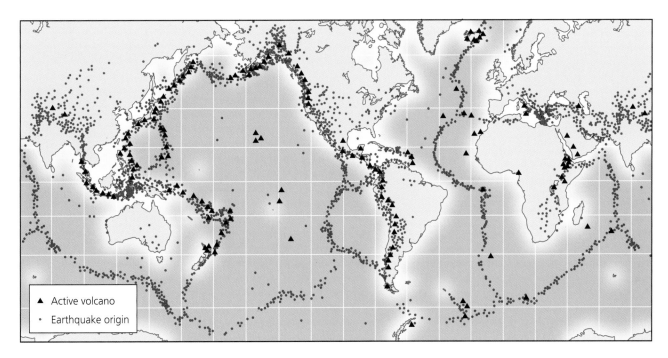

Figure 15.1 Global distribution of recent earthquakes and active volcanoes.

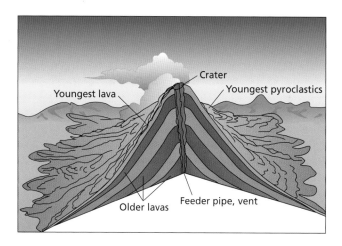

Figure 15.2 Simplified cutaway view of a composite volcano, showing a sequence of lavas interbedded with compacted pyroclastics.

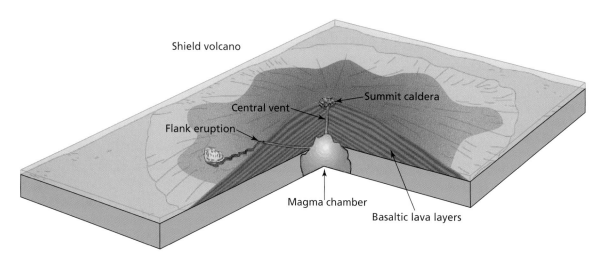

Figure 15.3 Simplified cutaway view of a shield volcano. Vertical scale is greatly exaggerated so that the mountain looks more steep-sided than it actually is. The base of this volcano extends over 320 km (200 mi); its height above the ocean floor is around 13 km (8 mi).

as well as in the Aleutian Islands, the Kamchatka peninsula, the Philippines, and throughout the islands of the southwest Pacific. While this is a particularly volcanically active region, volcanoes can be found in any part of the world where tectonic activity is present.

There are three basic types of volcanoes. The first is the *shield volcano* (Fig. 15.3). These are the least violent of the three types and are formed of layer upon layer of basaltic lava flows. The volcanoes of Mauna Loa and Kilauea in Hawaii are shield volcanoes. The second type is the *cinder cone volcano*. These are made of layers of cinders, a combination of ash and pyroclastic materials. While these are more vigorous than shield volcanoes, they are still less violent than the most dangerous of the three types, the composite or, as they are also called, *stratovolcanoes*. These are the most powerful, destructive, and potentially deadly volcanoes. Examples of composite volcanoes are Mount St. Helens in Washington state and Mt. Pinatubo in the Philippines. These were very violent eruptions that took a substantial number of lives and caused tremendous property damage. Perhaps the most famous eruption of all time was that of a stratovolcano, Mt. Vesuvius, near present-day Naples, Italy. In 79 A.D., it destroyed and buried (and preserved) the Roman towns of Pompeii and Herculaneum.

As magma emerges onto the surface, it loses most of the gasses it contains and becomes lava. There are two types of lava; the type that is more fluid and will flow is called *pahoehoe*. The other type, which is less fluid and is blocky shaped due to its mineral composition, is called *aa*. There are also other materials associated with volcanoes, such as *lahar*. Lahar is a mixture of ash, mud, and water, which, as it flows downslope and through river valleys, demolishes and buries anything in its path. Another type is called *nuée ardente*, which means "glowing cloud." This is a mass of glowing cinders and pyroclastic materials that combine in a cloud of extremely high temperatures that emerge from an erupting volcano and descend downslope, literally incinerating everything in its path. This is what destroyed the city of St. Pierre, Martinique, and killed thousands of people during the eruption of Mount Pelée in 1902. It has recently been determined that this was the cause of death of more than three hundred people discovered in the remains of the Roman town of Herculaneum as a result of the 79 A.D. eruption.

Not all volcanoes are found on plate boundaries. Some form within the plates away from the boundaries, where columns of fluid, extremely hot rock rise up through the mantle. These columns are called *plumes* and remain stationary in the mantle while the plates move over them. Near the top of these plumes magma forms and flows upward through the crust, forming a volcano on the surface. These are known as "hot spots" and are found primarily in oceanic plates but sometimes form on continental plates as well. When an oceanic plate moves over a plume, it can lead to the formation of volcanic island chains, as each volcano formed rises above sea level. The Hawaiian Islands are a volcanic island chain formed from a hot spot. Island chains that form over hot spots should not be confused with volcanic island arcs, which form along subduction complexes. The Aleutian Islands and the Mariana Islands are volcanic island arcs.

Figure 15.4 A huge cloud of volcanic ash and gas rises above Mt. Pinatubo in the Philippines on June 12, 1991. Three days later, an explosive eruption, one of the largest in the past century, spread a layer of ash over a vast area around the mountain with environmental as well as political consequences: a dust cloud surrounded Earth and reduced global warming, and a nearby U.S. military base on Pinatubo's island (Luzon) was damaged beyond repair.

A

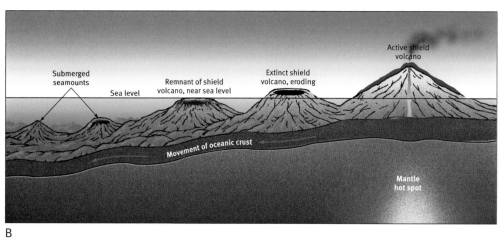

B

Figure 15.5 Ocean floor of the northwestern Pacific, dominated by the Emperor Seamounts and the Hawai'ian chain (A). Volcanic chain formed by the sea floor moving over a geologic hot spot (B). These volcanic landforms become progressively older toward the left.

1) Briefly describe the difference between a shield volcano and a cinder cone volcano.

_____

_____

_____

2) What is a composite volcano and how is it different from shield volcanoes and cinder cone volcanoes?

_____

_____

_____

3) Where is the largest concentration of the world's volcanoes found and why?

_____

_____

_____

4) Explain why there is so much volcanic activity in Iceland.

_____

_____

_____

5) What is lahar and what makes it so destructive?

_____

_____

_____

6) Using your world map and the map depicting plate boundaries and volcanoes, locate the following volcanoes and determine whether or not they are on a plate boundary and, if so, what kind of boundary:

    a. Mount Vesuvius

_____

    b. Mount Pinatubo

_____

    c. Mount St. Helens

_____

    d. Mauna Loa

_____

7) What are the two most common types of lava and how are they different?

_____

_____

_____

8) What U.S. volcano erupted on May 18, 1980? In which state was it located and what mountain range was it part of? What type of volcano was it?

_____

_____

_____

9) What is a hot spot? Explain the processes associated with it.

_____

_____

_____

10) What is the difference between a volcanic island chain and a volcanic island arc? Give an example of each.

_____

_____

_____

CRITICAL THINKING QUESTIONS

1) What impact can volcanoes have on climate?

_____

_____

_____

2) On air quality?

_____

_____

_____

3) On air travel?

_____

_____

_____

4) Why are geologists currently concerned about Yellowstone National Park?

_____

_____

_____

EARTHQUAKES

GOALS OF THIS LAB:

To understand earthquakes and the processes that cause them.

MATERIALS REQUIRED:

Textbook and a geopolitical map of the world

Earthquakes are one of the most catastrophic of geologic events. They can level cities, cause seas to rise, and alter coastlines. While they may seem random in their occurrence, they can be somewhat predictable, at least to a general degree, if we understand the geologic processes that produce them. Once again, as we saw in Lab 15 with volcanoes, earthquakes are closely integrated with plate tectonics. Technically, any sudden movement, trembling, or shaking of the Earth's crust constitutes an earthquake. While we tend to think of the infrequent violent movement that results in significant physical damage as an earthquake, most earthquakes are small motions. Many are imperceptible to human senses and occur virtually all over the world on a daily basis. We would not know of their existence if not for technology that can measure even the slightest motion of the Earth's crust.

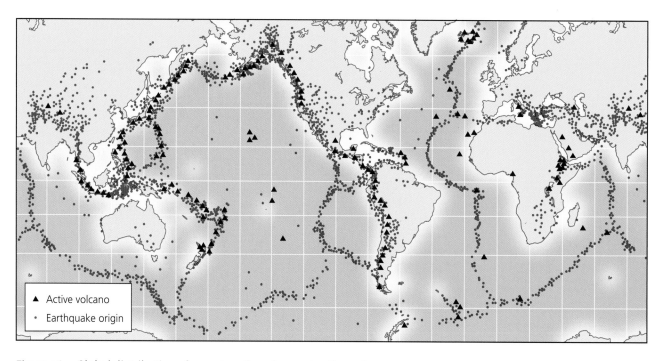

Figure 16.1 Global distribution of recent earthquakes and active volcanoes.

The technology that is used to alert us to the motion of the Earth's crust when we are not aware of it taking place, and to give us a way of gauging the power of earthquakes, is the *seismograph*. Seismographs measure the force or intensity of an earthquake. Geologists who specialize in studying earthquakes are seismologists. By studying where and how earthquakes occur, seismologists are working toward the goal of accurate earthquake prediction, which could potentially save many lives. Earthquakes are caused by the release of energy that builds in the Earth's tectonic plates as they shift and move. They can occur at the Earth's surface, in the plates, and even in the upper mantle. When they do occur, the release of energy sends seismic waves out in every direction from the point at which the earthquake begins. This point in the crust is called the *focus*, and the point above it at the Earth's surface is called the *epicenter*. Tectonic stresses can fracture the Earth's crust, resulting in a *fault*. When an earthquake takes place on or near a fault, the release of energy can cause the fault to increase in size or cause the fault to shift suddenly in response to the pressures exerted on it. A good example of a fault that is under pressure and constantly undergoing correction from intense pressures is the San Andreas Fault in California.

There are two methods of measuring the strength of an earthquake: by its magnitude and by its intensity. While these may seem similar, they are different. Magnitude is the measurement of the energy released when an earthquake occurs. Intensity is the severity of the shaking of the Earth's surface at a particular location. The original scale used for measuring magnitude was the Richter scale, named for the geologist who developed the original version in the 1930s. Today, we use a modernized version of the original scale called the *moment magnitude scale*, which is very similar to the Richter scale but rates earthquakes somewhat differently. The important

thing to remember about both the Richter and moment magnitude scales is that they are logarithmic. For example, as you move up the

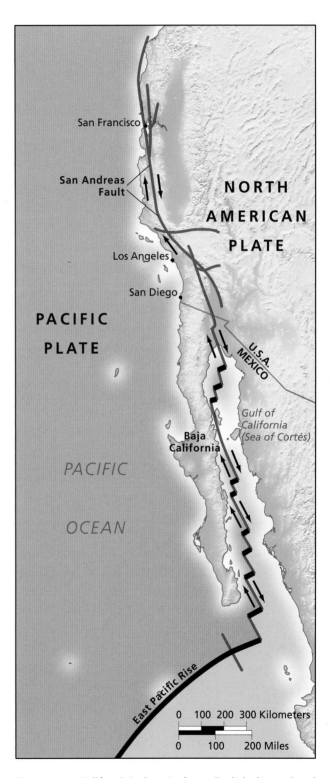

Figure 16.2 California's San Andreas Fault in its regional context. This fault separates the Pacific Plate from the North American Plate, where the two plates slide past each other at a transform boundary.

scale from a magnitude 6 to a magnitude 7, the increase to a magnitude 7 is *ten* times the magnitude of a 6. One can see that as you go even higher on the scale to the 8 and 9 range, the magnitude can be catastrophic. The earthquake that struck the Indonesian region on December 26, 2004, occurred on the sea floor and was a 9.1 on the moment magnitude scale. The earthquake that struck the northern end of the island of Honshu in Japan in 2011 was also at least a 9. Both of these earthquakes caused tremendous amounts of physical damage to the respective regions, including a significant number of casualties, particularly the Indonesian quake, which resulted in over 200,000 deaths throughout the affected region.

The system used for measuring intensity is the *modified Mercalli scale*, named after the Italian geologist who developed it at the beginning of the twentieth century. It categorizes earthquakes based on intensity by using a scale of Roman numerals, I through XII. Table 33.2 in the fifth edition of *Physical Geography* has a chart explaining how this scale works.

Besides the obvious destruction that a large earthquake can cause, there are other forces associated with earthquakes that can be more devastating than the earthquake itself. The most significant of these are *tsunamis*. Tsunamis are massive surges of the oceans that are the result of large earthquakes on the sea floor. The large loss of life in conjunction with the Indonesian earthquake of 2004 was the result of tsunamis that radiated out from the epicenter of the earthquake off the coast of Sumatra. While the earthquake in Sumatra occurred in the Indian Ocean, the majority of major earthquakes take place around the circum-Pacific

TABLE 16.1 Noteworthy Earthquakes of the Twentieth and Twenty-First Centuries

Year	Place	Moment Magnitude	Estimated Death Toll
1906	San Francisco, California	7.8	3,000
1908	Messina, Italy	7.2	72,000
1920	Ningxia, China	7.8	200,000
1923	Tokyo-Yokohama, Japan	7.9	143,000
1960	Temuco-Valdivia, Chile	9.5	1,655
1964	Prince William Sound, Alaska	9.2	131
1970	Chimbote, Peru	7.9	70,000
1976	Tangshan, China	7.5	242,000*
1976	Guatemala	7.5	23,000
1985	West-Central Mexico	8.0	9,500
1988	Armenia, (then) U.S.S.R.	6.8	25,000
1990	Northwestern Iran	7.4	50,000+
2001	Gujarat, India	7.6	20,000
2004	Sumatra, Indonesia	9.1	228,000
2005	Northern Pakistan	7.6	86,000
2008	Sichuan, China	7.9	87,600
2010	Port-au-Prince, Haiti	7.0	316,000
2011	Northern Japan	9.0	20,986

*Official death toll. Reliable reports persist that as many as 700,000 died in this earthquake. Moment magnitudes listed are those calculated by the U.S. Geological Survey.

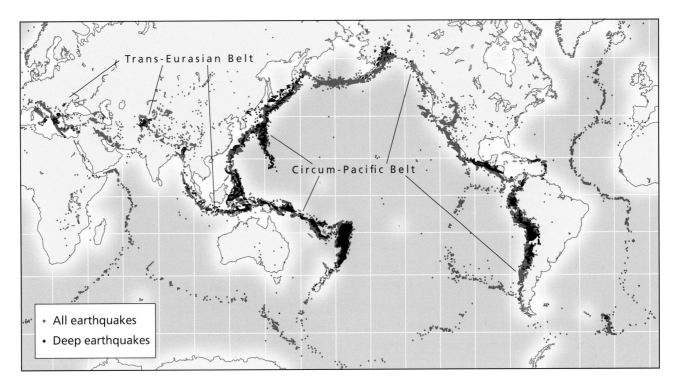

Figure 16.3 Global distribution of recent earthquakes. The deep earthquakes, shown by black dots, originated more than 100 km (63 mi) below the surface.

Figure 16.4 A ferocious earthquake (magnitude 9.2) devastated parts of Anchorage, Alaska, on March 27, 1964. This is Fourth Avenue the day after; it shows only part of the destruction that extended from the harbor to the interior.

belt (refer back to what we learned in the exercises on volcanoes and the "Ring of Fire" in Lab 15). The earthquake that struck Japan in 2011 is indicative of this region's instability. Both the earthquake in Japan and the earthquake in Indonesia were similar in that they both occurred near massive subduction zones. Subduction zone earthquakes can be very violent and almost always trigger tsunamis.

As discussed in the exercise on plate tectonics in Lab 14, these subduction zones are where a continental plate is overriding an oceanic plate and forcing it down into the Earth's mantle. Tremendous forces are at work as this process is taking place, and sudden, catastrophic releases of energy can result in earthquakes like those seen in Indonesia, Japan, Alaska, and Chile.

Not all earthquakes occur along plate boundaries. In fact, many occur within the plates far from the nearest boundary. A good example of this is the great earthquake that struck southeastern Missouri in 1811 near the community of New Madrid. Estimated at a magnitude of 8.5, it forced the Mississippi River to flow backward, causing flooding and forming lakes that still exist in the region today. Had the cities of St. Louis and Memphis existed in 1811 as major urban areas, there would have been massive destruction and great loss of life. It should be noted that the forces that caused the 1811 earthquake are still present today, and the danger of a major earthquake in this region continues.

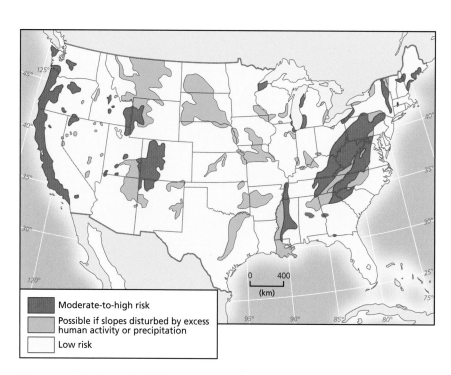

Figure 16.5 Earthquake risk in the conterminous United States.

1) What machine do geologists use to measure earthquake intensity and how does it work?

_____

_____

_____

2) What is the difference between the modified Mercalli scale and the Richter scale?

_____

_____

_____

3) What is the difference between the focus and the epicenter of an earthquake?

_____

_____

_____

4) Name two regions of high earthquake activity.

_____

_____

_____

5) What is a tsunami?

_____

_____

_____

6) Name two recent large earthquakes that resulted in extremely large numbers of casualties and explain why they occurred where they did.

_____

_____

_____

7) Explain why a strong earthquake occurred near New Madrid, Missouri, in 1811.

_____

_____

_____

8) What is the difference between attenuation and amplification?

_____

_____

_____

9) What is a fault, what role does it play in earthquake activity, and how is it different from a plate boundary?

_____

_____

_____

10) Using Figure 16.5, name two potentially strong intraplate earthquake zones in the United States.

_____

_____

_____

CRITICAL THINKING QUESTIONS

1) Why do places such as southern China, India, Turkey, and California seem to constantly experience earthquake activity? Give specifics for each place in your answer.

_____

_____

_____

2) Explain the circumstances surrounding the massive earthquake near Indonesia in December 2004 and why there was such a large loss of life. Could anything have been done to reduce the number of casualties?

_____

_____

_____

WEATHERING PROCESSES

GOALS OF THIS LAB:
>To understand the processes that break down rocks on the Earth's surface.

MATERIALS NEEDED FOR THIS LAB:
>Textbook

As we have seen in previous exercises on plate tectonics, the Earth is constantly pushing up mountains, volcanoes, and other features. As soon as these form, forces of nature begin to attack and destroy them. This destructive process is called *weathering*. Physical geography defines weathering as the in-place destruction of Earth's materials by physical and chemical forces. There are two basic types of weathering processes at work on Earth, *physical* and *chemical.*

Physical weathering, which is sometimes also called *mechanical* weathering, includes forces that break the bonds that hold rocks and minerals together. Gravity is a major force in physical weathering, as is water, wind, and glacial ice. There are many types of physical weathering processes. Some of these are:

1. *Frost wedging.* Every time it rains in cooler weather, some of the water seeps into fractures in rocks and rock formations. As nightfall approaches and temperatures drop, that water freezes. One of the interesting properties of water is that it is one of the few substances that expand when it freezes,

unlike most materials, which contract. The result of the expansion of this ice in the cracks is that it forces the pieces farther apart, loosening the material until gravity takes over and it falls apart. This debris drops downslope and builds up at the base. One of the important things about weathering is that as rocks break down, more surface area is exposed and the weathering process is accelerated.

2. *Exfoliation.* Also known as *sheeting.* Exposure to the elements causes certain rocks and rock formations to peel off in thin layers, somewhat like peeling layers from an onion. Large formations that weather in this way are called *exfoliation domes.* A prominent example of an exfoliation dome is Stone Mountain, Georgia.

3. *Thermal expansion.* This is a process that primarily occurs when rocks are rapidly heated and cooled. As rocks absorb heat, they expand, and when they cool down, they contract. When these temperature variations occur quickly, the stress from the constant

expansion and contraction causes rock to break down. This can happen as a result of a forest fire sweeping over a rock outcrop. Because of the lack of humidity in arid regions of the world, the atmosphere does not trap and hold heat. As temperatures get extremely hot during the day, rocks absorb the heat, causing them to expand. This heat is released into the atmosphere during the night, and with no humidity, temperatures can fall below freezing, causing rocks to contract. While it is the subject of debate among some scientists, recent studies have shown it is possible that this constant expansion and contraction causes rocks to break down.

4. *Organic activity.* This is sometimes considered an additional class along with physical and chemical and is referred to as *biological* weathering. This type of weathering is the result of organisms contributing to the breakdown of rocks and rock formations. Examples of organic activity would be tree roots, animal burrows, and human actions.

Figure 17.1 This massive granite boulder in Joshua Tree National Park, California, has been split in two along a joint plane by temperature fluctuations and the wedging effect of moisture. So far neither part of the boulder has been dislodged from the place where it was wedged apart, but gravity may act differently on the two masses as time goes on.

Chemical weathering is a chemical reaction that causes the bonds that hold rocks and minerals together to uncouple. This compromises the structural integrity of the crystal lattices that hold them together, particularly in silicate rocks, causing them to break down. One of the primary agents of chemical weathering is water. While we generally think of water as being a neutral substance, as the hydrologic cycle takes place water molecules react with carbon dioxide in the atmosphere and produce a weak solution of carbonic acid. Carbonate rocks are particularly reactive to even a weak acid, limestone being the most abundant of these. While some are particularly susceptible to chemical weathering, all rocks are subject to it. As with physical weathering, there are several chemical weathering processes. Some of these are:

1. *Solution.* The process whereby minerals and rocks dissolve.
2. *Hydrolysis.* This occurs when water molecules separate to release hydrogen ions, which then react with minerals and break them down. Hydrolysis usually plays an essential role in the weathering of silicate minerals, which causes the breakdown of rocks such as granite, which are composed of silicates. This leaves behind only the most resistant silicate minerals such as quartz. The further breakdown of feldspars that are released through hydrolysis contributes to the formation of the clay minerals.
3. *Oxidation.* This process occurs when oxygen reacts with metallic elements to form oxides or, if water is present, hydroxides. This is the same process as rusting that occurs on materials made of iron. Rocks that are undergoing oxidation will often have a yellow or orange dusting on them.

One result of chemical weathering processes is called *Karst topography.* It occurs when the rock layer below the surface is carbonate, usually limestone, and the process of solution dissolves the underlying rock over time. This will result in a multitude of surface and subterranean features forming that are somewhat unique. These areas can be quite large, sometimes covering substantial portions of entire states. There are large Karst regions found in Kentucky and Florida in the United States as well as China and other locations throughout the world.

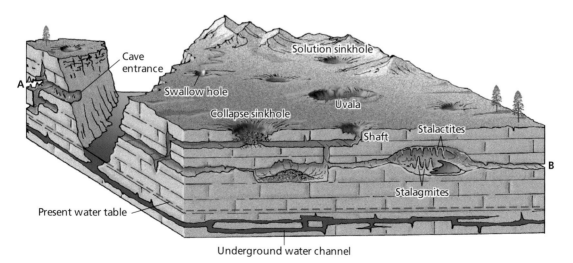

Figure 17.2 Surface and underground features of temperate karst. When the groundwater table was higher, the solution features at levels **A** and **B** formed. Now the water table is lower and solution proceeds. A new cave network will form when the next drop in the water table occurs.

TABLE 17.1 Reactions Involved in Chemical Weathering

Carbonation and Carbonate Mineral Weathering

1. CO_2 + H_2O → H_2CO_3 → H^+ + HCO_3^-

    carbon dioxide (gas)   water   carbonic acid   hydrogen ion   bicarbonate ion

2. $CaCO_3$ + H^+ → Ca^{2+} + HCO_3^-

    calcite (a carbonate, solid)   hydrogen ion   calcium ion   bicarbonate ion

Hydrolysis and Silicate Mineral Weathering (a simple example; more complicated silicate weathering reactions also involve formation of clays or other secondary minerals)

1. $4H_2O$ → $4H^+$ + $4OH^-$

    water   hydrogen ion   hydroxide ion

2. $4H^+$ + Mg_2SiO_4 → $2Mg^{2+}$ + H_4SiO_4

    hydrogen ions   olivine (a silicate, solid)   magnesium ions   silicic acid (dissolved)

Oxidation (example of pyrite)

FeS_2 + $2H_2O$ + $4O_2$ → $FeO(OH)$ + $4H^+$ + $2SO_4^{2-}$

pyrite (a sulfide mineral)   water   oxygen   goethite (an oxide mineral)   hydrogen ions   sulfate ions

EXERCISES

1) Briefly describe the differences between physical and chemical weathering.

_____

_____

_____

2) What is talus?

_____

_____

_____

3) What does *felsenmeer* mean?

_____

_____

_____

4) Explain the process of carbonation.

_____

_____

_____

5) Explain the process of hydrolysis.

_____

_____

_____

6) What is kaolinite and where does it come from?

_____

_____

_____

7) What are some forms of biological weathering?

_____

_____

_____

8) What is a good example of an exfoliation dome?

_____

_____

_____

9) What is thermal expansion and how does it work?

_____

_____

_____

10) Explain the process of oxidation and its results.

_____

_____

_____

CRITICAL THINKING QUESTIONS

1) Explain what Karst topography is and what role chemical weathering plays in it.

_____

_____

_____

2) How does Karst topography impact human activities? Give specific examples in your answer.

_____

_____

_____

WATER

GOALS OF THIS LAB:
To understand the importance of water as a natural resource and its function as a geomorphological force.

MATERIALS NEEDED FOR THIS LAB:
Textbook and a geopolitical map of the United States

Of all the resources on Earth, perhaps none is more overlooked and taken for granted than water. While most of the Earth's surface is covered with water (71 percent), it is saline and can only be made useful to humans through an expensive process called *desalinization*. In fact, 97 percent of the water on Earth is saline. Only 3 percent is freshwater, which is necessary for drinking, agriculture, and other applications. An additional obstacle is that most of this freshwater is locked up in glacial ice and polar ice-caps and is not readily accessible. Water is recycled through a process called the *hydrologic cycle*. This begins with evaporation from surface water sources. Water vapor rises into the atmosphere where condensation occurs, resulting in the formation of clouds. Precipitation then falls back to the Earth's surface. It is water that makes the Earth unique in that it is found here in all three states of matter, as a liquid, a solid, and a gas. As far as we know, Earth is the only place where it exists in all three states.

As water falls in the form of precipitation to the Earth's surface, it does so because of the force of gravity. When it reaches the surface, it is still under the influence of gravity and will move to the lowest possible elevation. As raindrops fall on the land surface areas of the Earth, they often land on soil. A certain amount of a soil's volume is actually air in the form of pore spaces in between the sand, silt, and clay particles and organic matter that compose it. Large pore spaces are *macropores*, and small pore spaces are *micropores*. Due to the effects of gravity, water will move downward through these pores. This movement is called *infiltration*. The area that comprises pore spaces that are normally filled with air, through which water can freely infiltrate, is the *zone of aeration*. As water moves downward, it will eventually reach a layer through which water cannot freely pass and will not move any further. Water cannot penetrate this layer, and it is said to be *impermeable*. An impermeable layer that prevents the further downward infiltration of water is called an *aquiclude*. Water will then accumulate in the pore spaces of the soil immediately above the aquiclude in what is called the *saturated zone*. Where the zone of aeration meets the saturated zone is called the *water table*. An accumulated reservoir of water below ground that can flow freely enough to supply a well or spring is called an *aquifer*.

Figure 18.1 Waterfall in the Scottish Highlands at Glen Coe, Scotland.

There are two basic types of aquifers, *confined* and *unconfined*. Unconfined aquifers are usually the saturated zone that rests on an aquiclude near the surface. Confined aquifers are deeper water-bearing strata that are trapped between two aquicludes. The entire surface constitutes the recharge area of an unconfined aquifer. The recharge area of a confined aquifer can be many miles away from the location where water is being withdrawn, as the water-bearing strata can only be recharged where erosion has exposed it at the surface. Water that accumulates below the surface in these ways is called *groundwater*.

Once pore spaces of a soil are filled with water, or saturated, the rate of infiltration slows and they cannot absorb the rainfall. Because this water cannot be absorbed into the ground, it now moves along the surface of the ground. This water is called *runoff*, also referred to as

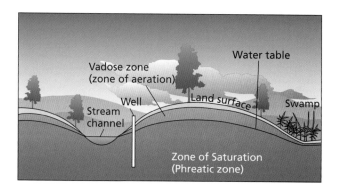

Figure 18.2 Two subsurface water-holding zones. The term *groundwater* applies to water lying below the water table—in the saturated zone.

running water. Any channelized flow of water is called a *stream*. A stream will find the lowest elevation and continue to flow down gradient until it reaches its *ultimate base level*, which is sea level. Most of the water that falls to the ground as precipitation will become running water and eventually will find its way back to the ocean, thereby completing the hydrologic

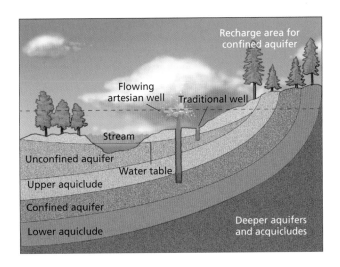

Figure 18.3 Aquifers, aquicludes, and their relationship to the table and wells. The dashed red line is the level water in the confined aquifer will rise in a well.

Figure 18.4 Sandbars formed by sediments in River Mawddach, near Barmouth, Wales.

cycle. Streams play a major role as an erosional force that sculpts the land surface of the Earth. The material that it erodes is transported by a stream in three ways. Material that is water soluble will go into solution and is called *dissolved load.* Material that is not water soluble but is small and light enough to be carried by the force of the moving water in suspension is called *suspended load.* Finally, there are some materials that are nonsoluble and too heavy to be carried in suspension but are subject to being swept along the bottom of a stream by rolling, sliding, or a process called *saltation.* Saltation is when materials are too heavy to be transported in suspension but can be bounced off the bottom and carried along in the water flowing above the streambed before settling and bouncing along again. Materials moved along the bottom of a stream are called *bed load*. The Mississippi River is an example of how much sediment can be carried by a large stream. If you have ever seen the Mississippi, you know that the water is a brown color due to the large sediment load it carries. The Mississippi annually transports roughly 750 million tons of sediments. About 500 million tons are in the form of suspended load, about 200 million tons are dissolved load, and the remainder is in bed load. Plainly, the majority of the Mississippi's sediment load is carried as suspended load.

Because of the erosive nature of running water, the sediments that are eroded from one place will be deposited somewhere else as depositional, or *fluvial*, features. One common depositional feature is the buildup of sediments in and along the stream channel itself; these are called *bars*. They are often composed of the heaviest sediments that settle out first, primarily sand. As erosion continues, a stream will begin to develop a winding course. Over time the stream will erode laterally, creating a flat area on both sides of its channel called a *floodplain*. When a stream has periods of discharge that are greater than the primary channel can contain, water will flow out over the floodplain until the discharge declines and the water recedes back into the primary channel once again. As the floodplain widens, the bends in the course of the stream become more exaggerated, creating extreme curves in its course, called *meanders*. Eventually, the narrow neck of land that separates these severe curves in the stream's course can erode through. This produces a *cutoff*, resulting in an abandoned channel called an *oxbow lake*. The course of the Mississippi River has a well-developed meander belt and is lined with oxbow lakes, as well as an extensive floodplain.

Where a river empties into the sea is called its *delta*. There are several types of deltas. The type of delta is often the result of how the stream

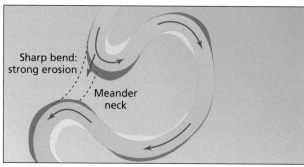

A

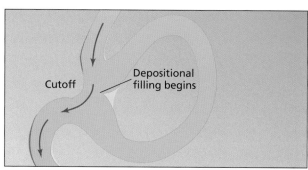

B

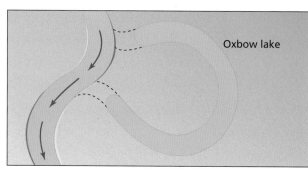

C

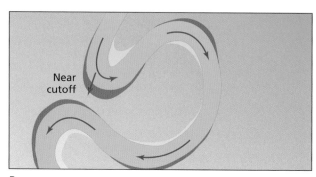

D

Figure 18.5 Process of meander cutoff, forming an oxbow lake.

enters the sea. Where a river empties, the velocity decreases and it can no longer carry some of the suspended load, and these sediments fall out. Over time they will begin to build up and fill the primary channel, causing the formation of distributaries that flow around the built-up materials. If the stream enters the sea in an area of shallow water, it will build a large area of marshes and bars with many distributaries, such as the Nile River delta. If it enters the sea where the water is deep, as does the Congo River, it will not build out very far, since most of the sediments will fall to the seafloor or be swept out by deep water currents. A good example of a stream that has built a considerable delta that has several large distributaries is that of the Mississippi River. This type of delta is called a *birdsfoot* delta and over time has built an extensive area of marshes and wetlands. This region is an important wintering area for waterfowl and a breeding ground for fish, amphibians, and reptiles. It can also act as an effective buffer to hurricanes when they strike the Gulf coast.

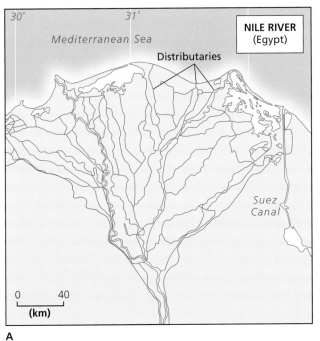

A

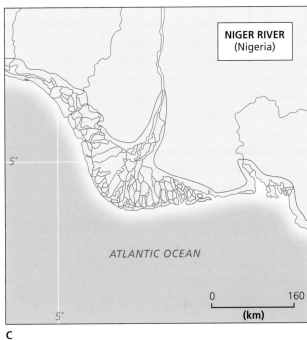

C

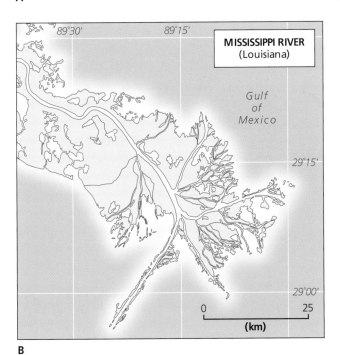

B

Figure 18.6 The spatial form of a delta depends on the quantity of sediment carried by the river, the configuration of the continental shelf beyond the river mouth, and the power of waves and currents in the sea. The Nile Delta (A) exhibits the classic triangular shape. The Mississippi Delta (B), exhibiting a birdsfoot shape, results from large quantities of sediment carried into the ocean too fast for extensive reworking by the local currents and waves. The Niger Delta (C) is shaped by strong waves and currents that sweep sediment along the coast.

1) Explain the differences between an unconfined and a confined aquifer.

_____

_____

_____

2) Using a map of the United States, examine the course of the Mississippi River from St. Louis to its mouth and describe its characteristics. Based on those characteristics, classify it by age.

_____

_____

_____

3) Name the stages of stream development and give the characteristics of each.

_____

_____

_____

4) What is infiltration?

_____

_____

_____

5) Explain the difference between a stream's capacity and its competence.

_____

_____

_____

6) Describe the different types of deltas and give an example of each.

_____

_____

_____

7) What is the difference between an artesian spring and a perched water table?

_____

_____

_____

8) What is a cone of depression? How could it affect other nearby wells?

_____

_____

_____

9) Name and describe the different types of drainage patterns for streams.

_____

_____

_____

10) Describe how an oxbow lake forms.

_____

_____

_____

11) What is an alluvial fan?

_____

_____

_____

CRITICAL THINKING QUESTIONS

1) Explain some ways pollution could possibly contaminate aquifer systems.

_____

_____

_____

2) What are some possible problems associated with the use of water in agriculture?

_____

_____

_____

Figure 18.7 View of Loch Ness north from Urquhart Castle, Scotland. Loch Ness formed in the Great Glen Fault, resulting in a deep lake that stretches almost 23 miles. As a result of its combined depth and length, it contains an extremely large volume of water.

GLACIERS

GOALS OF THIS LAB:

To understand glaciers, their formation, and their impact on the geomorphology of the Earth.

MATERIALS NEEDED FOR THIS LAB:

Textbook and a geopolitical map of the world

Most of the freshwater on Earth exists as ice. Significant portions of the Earth's surface are covered with ice, some of it over water, such as the north polar ice cap. *Glaciers* are sheets of ice that form on land surfaces and are in motion.

There are several types of glaciers, with continental and mountain glaciers the most common. Both begin with the same process: the recrystallization of partially melted snow into a granular form of ice called *firn*. Over time, accumulated firn will compact, eventually resulting in the formation of dense glacial ice. Over an extended period of time a glacier will form that can grow to be thousands of feet thick. In the case of continental glaciers, they can be thousands of square miles in size as well. Continental glaciers, as their name implies, can grow large enough to almost cover entire continents. They form during an ice age, when, due to a

Figure 19.1 The Tonsina Glacier in Alaska flows down a mountain valley from the zone of ice accumulation at higher elevations. Loss of ice from the lower part of the glacier is evident from the meltwater flowing out of it at the lower left. The bare rock near the lower end of the glacier was covered by ice in the recent past, but now the glacier is retreating.

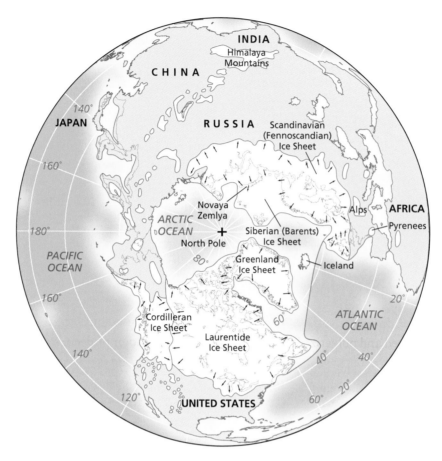

Figure 19.2 Late Cenozoic continental ice sheets and glaciers of the Northern Hemisphere, shown at the maximum size they reached during the last glacial period (not all reached this size at the same time). Arrows indicate the general direction of ice flow. Today's coastlines are shown as dashed lines. The Pleistocene coastlines on the map developed when the global sea level was about 100 m (330 ft) lower than at present. The pack ice covering the Arctic Ocean at that time extended much farther south, reaching the North Atlantic.

drop in global temperatures, much more precipitation falls in frozen form and collects on the continents. This is why as glaciation increases, sea level drops. There are currently only two continental glaciers on Earth, the Antarctic and Greenland ice sheets. The Antarctic, the larger one, is dome shaped and covers much of the continent of Antarctica. Because it is in motion, flowing from the center of the continent toward the margins, it has extended over water. When a glacier spreads out over water, that portion of the glacier is called an *ice shelf*. When these shelves form and extend out into the sea, they become brittle and fracture. Eventually chunks of ice will break

off in a process called *calving*, producing icebergs. During the most recent ice age, the Pleistocene Epoch, much of North America was repeatedly covered by a massive continental glacier, which at times reached as far south as St. Louis, Missouri, and Cincinnati, Ohio. After each expansion of the ice, the climate warmed and this glacier retreated and eventually disappeared. This large continental glacier, called the Laurentide Ice Sheet, substantially contributed to the formation of the Great Lakes.

Mountain glaciers are different than continental glaciers in that they occur at high elevations, specifically mountain valleys. Altitude plays an important role in the formation of mountain glaciers. High in the mountains where ridges and mountain peaks come together is often found a basin called a *cirque*. As snow falls and accumulates in a cirque it undergoes the process described previously producing firn, which leads to the formation of glacial ice. As the glacier grows, it begins to slowly move down the valley under the influence of gravity. Glaciers can move in one of two ways, by *glacial creep* or *glacial sliding*. The upper portion of a mountain glacier is called the *zone of accumulation*. The lower part is called the *zone of ablation*. As an ice sheet moves, it can develop fractures on its face called *crevasses*. When it moves through sliding, it can produce many depositional features made up of crushed and pulverized rock material called *glacial till*. Accumulated debris, which consists of rock, glacial till, and soil, forms deposits called *moraines* along the sides,

in front, and underneath. Those that form along the sides are *lateral moraines*. A ridge of glacial debris that accumulates along the terminus of a glacier is a *terminal moraine*. When a glacier recedes in stages and forms new terminal moraines behind the previous ones, they are called *recessional moraines*. Where two mountain glaciers merge into one larger glacier, two of the lateral moraines will also merge, forming a moraine down the center of the larger combined glacier called a *medial moraine*.

When a glacier recedes, or melts, there are other features left behind. Sometimes large chunks of the ice become embedded in the surface, which will eventually melt, resulting in the formation of deep, somewhat rounded lakes called *kettle lakes*. Mounds of debris that collect underneath a glacier and are steep sided all around are called *kames*. Kames usually

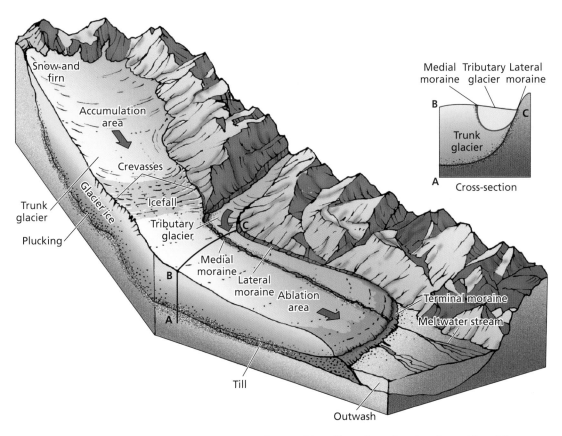

Figure 19.3 Cutaway view of a valley glacier and one of its tributaries, showing depositional features. Note the positions of the lateral, medial, and terminal moraines as well as outwash down the valley from the glacier.

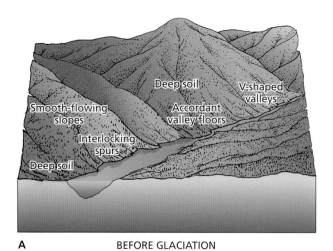

A   BEFORE GLACIATION

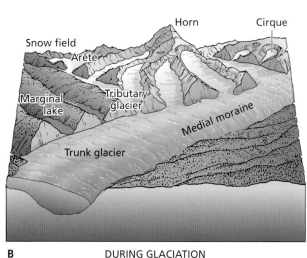

B   DURING GLACIATION

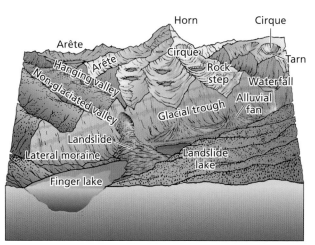

C   AFTER GLACIATION

Figure 19.4 Transformation of a mountain landscape by alpine glaciation (after William Morris Davis). Note how the initial rounded ridges and peaks are sculpted into a much sharper-edged topography by frost wedging and glacial erosion.

occur as solitary features. Mounds of debris that are teardrop shaped and occur in groups or fields are called *drumlins*. Occasionally, as temperatures warm, a mountain glacier will melt completely. Often when this occurs, in the cirque where the glacier originated, a lake will form called a *tarn*.

Any of these depositional and erosional features may be found wherever glaciers have occurred during the recent geologic past. Minnesota is called the "land of 10,000 lakes" because it has been underneath glaciers on several occasions. Each time a glacier melted it left behind scars that filled with runoff water, forming lakes. The state has many other glacial features as well as its famous lakes.

1) What is a cirque?

_____

_____

_____

2) What are pluvial lakes and how do they form?

_____

_____

_____

3) Explain how continental glaciers form and give present-day examples.

_____

_____

_____

4) Explain how mountain glaciers form.

_____

_____

_____

5) Using your world map, find three locations where you would find mountain glaciers and explain why.

_____

_____

_____

6) What are moraines? Name four types and explain how each is formed.

_____

_____

_____

7) What are fjords? Using your world map, give two locations where fjords are found and explain why.

_____

_____

_____

8) Name the two types of glacial movement and explain each.

_____

_____

_____

9) Explain the development of a hanging valley.

_____

_____

_____

10) Explain the differences between a kame and a drumlin and sketch an example of each.

_____

_____

_____

CRITICAL THINKING QUESTIONS

1) Explain the possible effects of global climate change, both warming and cooling, on glaciers and glaciation.

_____

_____

_____

2) How could those effects impact human populations?

_____

_____

_____

Figure 19.5 Glaciated terrain in the Cassiar Mountains of British Columbia, Canada. Note the tarn at left in the picture.

AEOLIAN PROCESSES

GOALS OF THIS LAB:
To understand the wind as a geomorphological agent.

REQUIRED ITEMS FOR THIS LAB:
Textbook, additional notebook paper, and a geopolitical map of the United States

As we have observed in previous chapters, erosion plays a major role in shaping the Earth's surface through the forces of running water and glaciers. The third major form of erosion is that of the wind. *Aeolian processes* are those that sculpt the surface by the force of the wind. Wind erosion is similar to water erosion in that the amount of materials that the wind can transport is related to its velocity and duration. Typically, aeolian processes take place in dry climates, primarily in deserts. This is due to the effects of water in holding soil particles in place and promoting the growth of vegetation in areas with higher precipitation amounts. Root systems of plants help hold soil particles in place, and vegetation also acts as a buffer to the force of wind near the surface where the majority of wind erosion occurs. Wind transports materials in suspension and along the surface in much the same way as streams move some of their bed load. In fact, the wind can bounce materials along the surface through the process of *saltation*, just as streams move some of their bed load in that way. The majority of particles moved by the wind are sand-sized grains, along with some silt and small amounts of clay particles. The force of the wind

smashing these particles against objects and landforms results in erosion through *abrasion*. Wind erosion is in essence sandblasting and can produce bizarre-shaped features over time. Time is an important variable, because wind erosion is very gradual and under most circumstances imperceptible to humans. Finer particles such as silt and clay can be lifted aloft by the wind in *suspension*, resulting in clouds of dust that can extend far above the Earth's surface. Sometimes heavier materials that cannot be lifted are moved along the floor of a desert in a process called *surface creep*. There are numerous features that are the result of aeolian processes. When materials are lifted away from the surface, *deflation* occurs. If deflation continues in the same area for an extended period of time, it can create a *deflation hollow*. Most deflation hollows are small, but occasionally they can be very large and quite deep. When fine material is constantly removed, leaving the larger, heavier particles behind, it can lead to the formation of *desert pavement*.

The erosion of materials from one place means they will be deposited somewhere else, leading to the formation of *depositional*

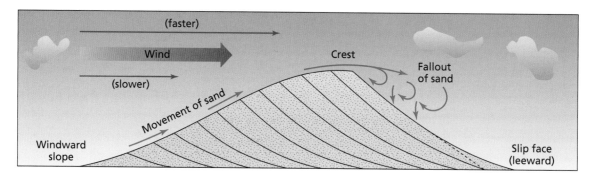

Figure 20.1 Cross-section of an active sand dune that is migrating from left to right. The lower-angle, windward slope is the erosional side of the dune, with surface sand grains pushed upward toward the crest. The steeper, leeward slip face is the depositional zone, where wind-driven sand is dropped by the wind and then slides downslope. The accumulation of sand grains on the advancing slip face produces strata inside the dune much like the foreset beds in a delta.

features. Perhaps the best-known aeolian deposits are called *dunes*. They are actually shaped by both erosion and deposition and are most common in areas called sand seas or *erg*. A dune has a defined structure. As the wind blows the sand and it builds, the windward side of the dune has a gradual slope. Because sand is subject to the effects of gravity, it will build to a point called the angle of repose, which is the steepest angle at which the loose sand is stable. When it exceeds its angle of repose, the downwind slope of the dune will collapse, once again correcting to the steepest angle at which it is stable. This is why the leeward side or steep side of a dune is called its *slip face*.

Dunes can form in several different shapes depending on the force and direction of the wind. The most recognized type of dune is called the *barchan*. Barchan dunes are crescent shaped, with their tips pointed in the downwind direction. While they are usually not particularly large, some can reach several hundred meters from tip to tip. The second type is the *transverse* dune. A transverse dune is a continuous ridge of sand that is formed at a right angle to the prevailing wind. Transverse dunes are sometimes confused with another type of dune, the *longitudinal*, because both are long ridges of sand. Longitudinal dunes are not perpendicular to the wind and

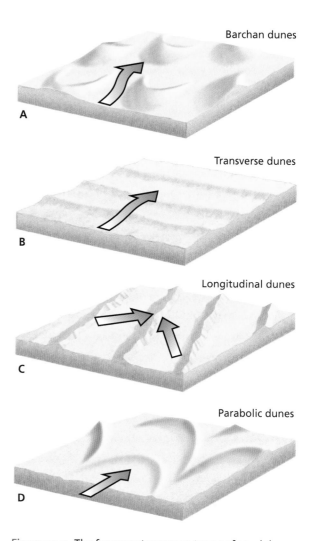

Figure 20.2 The four most common types of sand dunes.

can actually run in different directions, indicative of changes in wind direction. They can occasionally be found running parallel to the direction of the wind. The next type, the *parabolic*, is similar to the barchan in its crescent shape, except its tips point into the prevailing wind. The arms of the crescent are usually stabilized through the growth of vegetation.

Another substance deposited by the wind is a fine, dust-like material called *loess*. Geologically, loess has a composition of about 85 percent silt, 10 percent clay, and 5 percent sand. It can be distributed over a large geographic area as a blanket deposit. There are large areas of loess in China, Europe, and the United States. An example is the loess bluff region of southwestern Mississippi along the east bank of the Mississippi River, where the cities of Vicksburg and Natchez are located. These soils can be very fertile for agricultural purposes, and because they compact so tightly due to their small individual grain size, they have proven to be productive for paleontologists and archaeologists in preserving important fossil remains and artifacts.

1) What are yardangs?

_____

_____

_____

2) In what ways does the wind move particles? Explain each method.

_____

_____

_____

3) What is desert pavement? Describe the two possible processes that produce it.

_____

_____

_____

4) On a piece of notebook paper sketch a dune and label its parts. Explain what the angle of repose is and how it affects dune development.

_____

_____

_____

5) Name and describe four types of dunes and draw a simple sketch of each. Be sure to indicate wind direction on each sketch. Use additional notebook paper.

_____

_____

_____

6) What is loess and how does it form? Name three areas that have significant deposits.

_____

_____

_____

CRITICAL THINKING QUESTION

Explain the causes of the Dust Bowl and discuss the three primary states where it took place, how they were impacted by it, and what changes in agriculture were instituted in order to prevent a similar disaster from occurring again.

_____

_____

_____

COASTAL PROCESSES

GOALS OF THIS LAB:

To understand the processes that control coastal geomorphology.

REQUIRED ITEMS FOR THIS LAB:

Textbook and a geopolitical map of the world

We may think of the beach as a fun place to vacation or a great place to live, but it is also a complex area where the land masses transition to the sea. This zone is the focal point of many geologic processes that are constantly altering Earth's coastal regions. Features are continuously being formed and destroyed through a variety of forces. What we often refer to as the coast is the *littoral zone*. The shore is where the land actually meets the water.

One force that is constantly shaping Earth's coasts is the action of waves that exert energy against the shoreline as they strike the coastal zones. Waves originate in the oceans and are

Figure 21.1 Waves and surf along the southern coast of England near Charmouth, Dorset.

driven by the force of the wind. Waves have a definite structure. As one looks at waves as they approach land, they appear to have an undulating motion as they move. The highest point of a wave is its *crest*, and as it dips downward, its lowest point is its *trough*. The vertical distance between the crest and the trough is the wave's *height*, and the distance between two crests across the trough is the *wave length*. As the wind moves waves across the open ocean toward the shore, they build momentum and energy. As they approach the shore and the water gets shallower, friction begins to slow the wave until its crest rolls over just off the beach, producing *breakers*. Where this turbulent water crashes along the beach is called *surf*. Water in sheet form that rushes up on the beach from the surf is called *swash*, and when it runs back down to the surf it is called *backwash*. This wave energy that constantly batters the shores is a major erosional force that builds, alters, and destroys coastlines.

Both erosional and depositional features are found along coastlines. As waves strike the coast, they cause erosion to take place, particularly along protrusions of the coastline, called *headlands*, that extend out into the sea. As waves contact the coast, they bend around headlands and crash into them with great force, concentrating the erosive effects of wave energy there. This leads to the formation of steep, wave-cut *sea cliffs*. As the process continues, new features appear as a result of this erosion. Sea caves can be eroded out of the sides of the headland. When caves form on opposite sides and eventually meet, or a cave erodes completely through the headland from one side to the other, a *sea arch* is formed. Once the top of a sea arch erodes away, an isolated remnant of the headland is now left standing

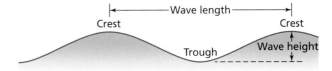

Figure 21.2 Wave height is the vertical distance between the wave crest and the wave trough. Wave length is the horizontal distance between two crests (or two troughs).

Figure 21.3 Wave-cut cliffs, sea stacks, and eroded headlands along the coast of Cornwall, England.

surrounded by water called a *sea stack*. Sometimes an opening will erode completely through a sea stack, resulting in another sea arch. These landforms and features are the result of erosion. As we have seen in other labs, if there is erosion one place, the eroded materials will accumulate somewhere else as depositional features. There are many depositional features associated with coastal regions. Because of the *longshore current*, eroded materials are swept along the shoreline, eventually accumulating and producing depositional features, such as *sand spits* and *baymouth bars*. This current is also responsible for the formation of *barrier islands* that usually lie just off of and parallel to the shoreline. Sometimes a sand spit will form that connects an island to the mainland, producing a *tombolo*.

Another aspect of the forces that impact coastlines is the normal, regular fluctuation of sea level. This rising and falling of the tides of the Earth's oceans is caused by the gravitational forces of the Sun and the Moon and their influence on those of the Earth. Each day there are periods of high tides and low tides. How high and how low they are is based on the positions of the Moon and Sun to the Earth. During certain phases, very high tides, called *spring tides*, will result.

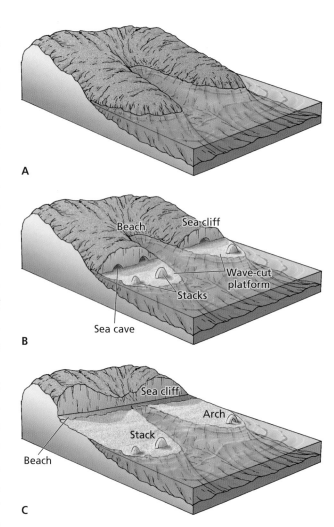

Figure 21.4 Straightening an embayed, indented coastline by wave erosion. When degradation is the dominant coastal process, a very different set of landforms develops (compare to the depositional landforms in Fig. 21.5).

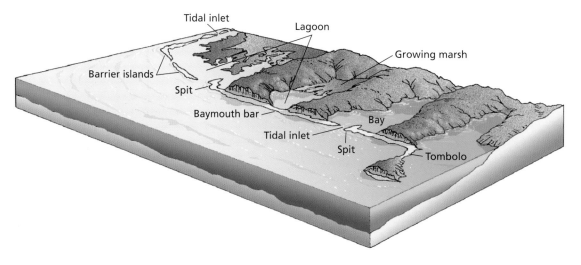

Figure 21.5 Common depositional landforms along a stretch of coastline.

During other phases, very low tides, called *neap tides*, will result. During spring tides, the distance between high tide and low tide will be the greatest. During neap tide periods, the difference between high and low tide will be the least. Tides play a role in coastal erosion and impact human activities in coastal regions—for example, ships and boats are often unable to enter harbors or rivers at periods of low tide.

Figure 21.6 Boats at low tide near the mouth of the River Exe, Topsham, Devon, England.

1) What are waves of oscillation?

_____

_____

_____

2) Explain the relationship between the Earth, the Moon, and the Sun that produces the tides.

_____

_____

_____

3) Explain how the longshore current works and what produces it.

_____

_____

_____

4) Name five erosional features found on coastlines and explain how each is formed.

_____

_____

_____

5) Name five depositional features found on coastlines and explain how each is formed.

_____

_____

_____

6) Explain what an emergent coast is.

_____

_____

_____

7) Explain what a submergent coast is.

_____

_____

_____

8) What is an atoll and how does it form?

_____

_____

_____

Figure 21.7 Eroded chalk cliffs along the English Channel near Dover, Kent, England.

9) Name three structures that are commonly built to protect property on coastlines and explain how they are designed to work.

_____

_____

_____

10) Using your map, find the five countries in the world with the most coastlines. Then find the three states in the United States with the most coastlines.

_____

_____

_____

CRITICAL THINKING QUESTIONS

1) List three densely populated urban coastal areas and explain in detail how they could be impacted physically and economically by a rise or fall in sea level.

_____

_____

_____

2) Describe the possible impacts on coastal areas by human settlement and activity.

_____

_____

_____

All figures appear in *Physical Geography*, 5th ed., unless otherwise noted †

Credits for Line Art and Tables

LAB 1

1.2: From Norman J. W. Thrower, *Maps and Man,* Prentice-Hall, 1972, p. 153. 1.4: After Arthur Robinson et al., *Elements of Cartography,* 5th rev. ed., 1984, p. 99, published by John Wiley & Sons. 1.5†: USGS. 1.6: After Edward A. Fernald and Donald J. Patton, *Water Resources Atlas of Florida,* 1984, pp. 26, 19, published by the Institute of Science and Public Affairs, Florida State University. 1.7†: USGS. 1.8†: Data courtesy of the North American Regional Reanalysis dataset, National Oceanic and Atmospheric Administration, Earth System Research Laboratory, Physical Sciences Division.

LAB 2

2.1, 2.2, 2.3, 2.4†: Maps courtesy of ESRI.

LAB 4

4.4: After M. I. Budyko, "The Heat Balance of the Earth," *Soviet Geography: Review and Translation,* vol. 3, no. 5, May 1962, pp. 7, 9, published by the American Geographic Society. 4.5: After J. G. Lockwood, *World Climatology: An Environmental Approach,* 1974, p. 43, published by Edward Arnold (Publishers) Limited, © Edward Arnold (Publishers) Ltd. 4.6†: Data courtesy of the National Centers for Environmental Prediction (NCEP) and the National Center for Atmospheric Research Reanalysis Dataset, National Oceanic and Atmospheric Administration, Earth System Research Laboratory, Physical Sciences Division. 4.7: Same as Fig. 4.4.

LAB 5

5.1: Data from K. Fennig, A. Andersson, S. Bakan, C. Klepp, M. Schroeder, 2012: Hamburg Ocean Atmosphere Parameters and Fluxes from Satellite Data—HOAPS 3.2—Monthly Means/6-Hourly Composites. Satellite Application Facility on Climate Monitoring. doi:10.5676/EUM_SAF_CM/ HOAPS/V001. 5.6†: Courtesy of the National Drought Mitigation Center at the University of Nebraska-Lincoln.

LAB 6

6.1: Adapted from Steven A. Ackerman and John A. Knox, *Meteorology: Understanding the Atmosphere,* 3rd ed., © (2012). Jones & Bartlett Learning, LLC, Sudbury, MA, p. 216. 6.3: Adapted from Robin McIlveen, *Fundamentals of Weather & Climate,* 2nd ed., © 2010. Oxford University Press, New York, NY, p. 111.

LAB 7

7.3: Adapted from Brian J. Skinner and Stephen C. Porter, *Physical Geography,* John Wiley & Sons, 1987, pp. 715, 711. 7.4: After W. D. Sellers, *Physical Climatology,* 1965, p. 84, published by The University of Chicago Press, © The University of Chicago Press. 7.5: Same source as Fig. 5.1.

LAB 8

8.2: Same source as Fig 5.1.

LAB 11

11.3: Adapted from Edward Aguado and James E. Burt, *Understanding Weather and Climate,* 6th ed., © 2013. Pearson Education, Inc., Upper Saddle River, NJ. 11.4: Adapted from Robert A. Rohde/Global Warming Art, globalwarmingart.com.

LAB 12

12.2: After USDA, Soil Conservation Service, n.d. 12.3: Data from Natural Resource Conservation Service (USDA).

LAB 14

14.3: Data from the U.S. Coast and Geodetic Survey and NOAA.

LAB 15

15.1: Same as Fig 14.3.

LAB 16

16.1: Same as Fig 14.3. 16.2: Same source as Fig. 3.13 in *Physical Geography* 5th ed., p 490. Table 16.1: Data courtesy of U.S. Geological Survey. 16.3: Data from the U.S. Coast and Geodetic Data Survey. 16.5: After NOAA, n.d.

LAB 18

18.2: Adapted from Brian J. Skinner and Stephen C. Porter, *The Dynamic Earth: An Introduction to Physical Geology*, John Wiley & Sons, 1989, p. 199. 18.3: From Léo F. Laporte, *Encounter With the Earth*, Canfield Press, p. 249, © 1975 by Léo F. Laporte.

LAB 19

19.2: Same source as Fig. 3.13 in *Physical Geography* 5th ed., p. 365. 19.3: Same source as Fig. 18.2. 19.4: After Armin K. Lobeck, *Geomorphology*, McGraw-Hill Book Company, 1932.

LAB 20

20.1: Same source as Fig. 3.13 in *Physical Geography* 5th ed., pp. 316, 322, 324.

LAB 21

21.2: After Keith Stowe, *Essentials of Ocean Science,* 1987, pp. 84, 87, and 124, published by John Wiley & Sons, copyright © 1987 by John Wiley & Sons, Inc. 21.5: Same source as Fig. 3.13 in *Physical Geography* 5th ed., pp. 393, 393.

Credits for Photographs

LAB 13

13.2, 13.4, 13.5, 13.6: Dalton W. Miller, Jr.

LAB 15

15.4: Dave Harlow, USGS. 15.5: Tibor G. Toth/National Geographic Image Collection.

LAB 16
16.4: © Bettmann/Corbis.

LAB 17
17.1: Bruce Molnia, USGS.

LAB 18
18.1, 18.4, 18.7: Dalton W. Miller, Jr.

LAB 19
19.1: Bruce Molnia, USGS. 18.2. 19.5: Dalton W. Miller, Jr.

LAB 21
21.1, 21.3, 21.6, 21.7: Dalton W. Miller, Jr.